YOU LOOK
LIKE A THING
AND
I LOVE YOU

YOU LOOK LIKE A THING AND I LOVE YOU

How Artificial Intelligence Works
and Why It's Making the World
a Weirder Place

Janelle Shane

VORACIOUS

Little, Brown and Company
New York ▪ *Boston* ▪ *London*

Voracious / Little, Brown and Company
Hachette Book Group
1290 Avenue of the Americas, New York, NY 10104
littlebrown.com

First Edition: November 2019

Voracious is an imprint of Little, Brown and Company, a division of Hachette Book Group, Inc. The Voracious name and logo are trademarks of Hachette Book Group, Inc.

The publisher is not responsible for websites (or their content) that are not owned by the publisher.

The Hachette Speakers Bureau provides a wide range of authors for speaking events. To find out more, go to hachettespeakersbureau.com or call (866) 376-6591.

All images created by the author, except the following: GANcats image on page 126 made available under Creative Commons BY-NC 4.0 license by NVIDIA Corporation; used here with permission. Quickdraw kangaroos images on page 135 made available under Creative Commons BY-4.0 license by Google. School plan images on page 146 © by Joel Simon; used with permission. Submarine images on page 200 © by Danny Karmon, Yoav Goldberg, and Daniel Zoran; used with permission. Skiers image on page 201 © by Andrew Ilyas, Logan Engstrom, Anish Athalye, and Jessy Lin; used with permission.

ISBN 978-0-316-52524-4
LCCN 2019945563

10 9 8 7 6 5 4 3 2

LSC-C

Printed in the United States of America

To my blog readers, who laughed at all the silliness, drew the weird creatures, spotted all the giraffes, and baked the neural net–generated cookies. Thank you for putting up with the horseradish brownies.

To my family, for being my biggest fans.

Contents

INTRODUCTION: AI is everywhere   1

CHAPTER 1: What is AI?   7

CHAPTER 2: AI is everywhere, but where is it exactly?   29

CHAPTER 3: How does it actually learn?   61

CHAPTER 4: It's trying!   109

CHAPTER 5: What are you really asking for?   140

CHAPTER 6: Hacking the Matrix, or AI finds a way   161

CHAPTER 7: Unfortunate shortcuts   168

CHAPTER 8: Is an AI brain like a human brain?   185

CHAPTER 9: Human bots (where can you *not* expect to see AI?)   209

CHAPTER 10: A human-AI partnership   219

CONCLUSION: Life among our artificial friends   234

Acknowledgments   237

Notes   239

Index   253

YOU LOOK
LIKE A THING
AND
I LOVE YOU

AI Is everywhere

Teaching an AI to flirt wasn't really my kind of project.

To be sure, I'd done a lot of weird AI projects already. On my blog, *AI Weirdness*, I'd trained an AI to come up with new names for cats — Mr. Tinkles and Retchion were some of its less successful ones — and asked an AI to generate new recipes, some of which would sometimes call for "peeled rosemary" or handfuls of broken glass. But teaching an AI to sweet-talk humans was something else entirely.

AI learns by example — in this case studying a list of existing pickup lines and using them to generate new ones. The problem: the training dataset on my computer screen was a list of pickup lines that I had collected from various internet sources, all of which were awful. They ranged from cheesy bad puns all the way to rude innuendo. Once I'd trained an AI to imitate them, it would be able to produce thousands more at the push of a button. And, like an impressionable child, it wouldn't know what it should and shouldn't imitate. The AI would start with a blank slate, knowing nothing about what pickup lines are (or even what English is), and would learn from the examples, doing its best to imitate all the patterns it found. Including the grossness. It wouldn't know any better.

I considered giving up on the project, but I had a blog post to write, and I had just spent an inconvenient amount of time collecting the example pickup lines. So I started the training. The AI began to look for patterns in the examples, inventing and testing rules that would help it predict which letters should appear in which order in a pickup line. Finally, training was over. With some trepidation, I asked the AI for some lines:

```
You must be a tringle? Cause you're the only thing here.
Hey baby, you're to be a key? Because I can bear your
    toot?
Are you a candle? Because you're so hot of the looks
    with you.
You're so beautiful that you say a bat on me and baby.
You look like a thing and I love you.
```

I was surprised and delighted. The AI's virtual brain (about the same complexity as a worm's) wasn't capable of picking up the subtleties of the dataset, including misogyny or cheesiness. It did its best with the patterns it had managed to glean…and arrived at a different, arguably better, solution to the problem of making a stranger smile.

Though to me its lines were a resounding success, the cluelessness of my AI partner may come as a surprise if your knowledge of AI comes from reading news headlines or science fiction. It's common to see companies claim that AIs are capable of judging the nuances of human language as well as or better than humans can, or that AIs will soon be able to replace humans in most jobs. AI will soon be everywhere, the press releases claim. And they're both right — and very wrong.

In fact, AI is *already* everywhere. It shapes your online experience, determining the ads you see and suggesting videos while detecting social media bots and malicious websites. Companies use AI-powered resume scanners to decide which candidates to interview, and they use AI to decide who should be approved for a loan. The AIs in self-driving cars have already driven millions of miles (with the occasional human rescue during moments of confusion). We've also put AI to work in our smartphones, recognizing our voice commands, autotagging faces in our photos, and even applying a video filter that makes it look like we have awesome bunny ears.

But we also know from experience that everyday AI is not flawless, not by a long shot. Ad delivery haunts our browsers with endless ads for boots we already bought. Spam filters let the occasional obvious scam through or filter out a crucial email at the most inopportune time.

As more of our daily lives are governed by algorithms, the quirks of AI are beginning to have consequences far beyond the merely inconvenient. Recommendation algorithms embedded in YouTube point people toward ever more polarizing content, traveling in a few short clicks from mainstream news to videos by hate groups and conspiracy theorists.[1] The algorithms that make decisions about parole, loans, and resume screening are not impartial but can be just as prejudiced as the humans they're supposed to replace — sometimes even more so. AI-powered surveillance can't be bribed, but it also can't raise moral objections to anything it's asked to do. It can also make mistakes when it's misused — or even when it's hacked. Researchers have discovered that something as seemingly insignificant as a small sticker can make an image recognition AI think a gun is a toaster, and a low-security fingerprint reader can be fooled more than 77 percent of the time with a single master fingerprint.

People often sell AI as more capable than it actually is, claiming that their AI can do things that are solidly in the realm of science fiction. Others advertise their AI as impartial even while its behavior is measurably biased. And often what people claim as AI performance is actually the work of humans behind the curtain. As consumers and citizens of this planet, we need to avoid being duped. We need to understand how our data is being used and understand what the AI we're using really is — and isn't.

On *AI Weirdness*, I spend my time doing fun experiments with AI. Sometimes this means giving AIs unusual things to imitate — like those pickup lines. Other times, I see if I can take them out of their comfort zones — like the time I showed an image recognition algorithm a picture of Darth Vader and simply asked it what it saw: it declared that Darth Vader was a tree and then proceeded to argue with me about it. From my experiments, I've found that even the most straightforward task can cause an AI to fail, as if you'd played a practical joke on it. But it turns out that pranking an AI — giving it a task and watching it flail — is a great way to learn about it.

In fact, as we'll see in this book, the inner workings of AI algorithms are often so strange and tangled that looking at an AI's output can be one of the only tools we have for discovering what it understood and what it got terribly wrong. When you ask an AI to draw a cat or write a joke, its mistakes are the same sorts of mistakes it makes when processing fingerprints or sorting medical images, except it's glaringly obvious that something's gone wrong when the cat has six legs and the joke has no punchline. Plus, it's really hilarious.

In the course of my attempts to take AIs out of their comfort zone and into ours, I've asked AIs to write the first line of a novel, recognize sheep in unusual places, write recipes, name guinea pigs, and generally be very weird. But from these experiments, you can learn a lot about what AI's good at and what it struggles to do — and what it likely won't be capable of doing in my lifetime or yours.

Here's what I've learned:

The Five Principles of AI Weirdness:

- The danger of AI is not that it's too smart but that it's not smart enough.
- AI has the approximate brainpower of a worm.
- AI does not really understand the problem you want it to solve.
- But: AI will do *exactly* what you tell it to. Or at least it will try its best.
- And AI will take the path of least resistance.

So let's enter the strange world of AI. We'll learn what AI is — and what it isn't. We'll learn what it's good at and where it's doomed to fail. We'll learn why the AIs of the future might look less like C-3PO than like a swarm of insects. We'll learn why a self-driving car would be a terrible getaway vehicle during a zombie apocalypse. We'll learn why you should never volunteer to test a sandwich-sorting AI, and we'll encounter walking AIs that would rather do anything but walk. And through it all we'll learn how AI works, how it thinks, and why it's making the world a weirder place.

CHAPTER 1

What Is AI?

Quick, AI! Calculate warp
coordinates for the Bal Panda
system!

Oops. Wrong kind of AI.
I'm just a guy in a robot suit.
This is awkward.

If it seems like AI is everywhere, it's partly because "artificial intelligence"
means lots of things, depending on whether you're reading science fiction
or selling a new app or doing academic research. When someone says they
have an AI-powered chatbot, should we expect it to have opinions and feel-
ings like the fictional C-3PO? Or is it just an algorithm that learned to guess
how humans are likely to respond to a given phrase? Or a spreadsheet that
matches words in your question with a library of preformulated answers?
Or an underpaid human who types all the answers from some remote loca-
tion? Or — even — a completely scripted conversation where human and AI

are reading human-written lines like characters in a play? Confusingly, at various times, all these have been referred to as AI.

For the purposes of this book, I'll use the term *AI* the way it's mostly used by programmers today: to refer to a particular style of computer program called a machine learning algorithm. This chart shows a bunch of the terms I'll be covering in this book and where they fall according to this definition.

Things called AI

Called AI in this book

Machine learning algorithms
Deep learning
Neural networks
Recurrent neural networks
Markov chains
Random forests
Genetic algorithms
Generative adversarial networks
Reinforcement learning
Predictive text
Magical sandwich sorters
Unfortunate murderbots

In this book, but not AI

Science fiction AIs
Rules-based programs
Humans in robot costumes
Robots reading scripts
Humans hired to pretend to be AIs
Sentient cockroaches
Phantom giraffes

woo0Ooo

oops

Everything that I'm calling "AI" in this book is also a machine learning algorithm — let's talk about what that is.

KNOCK, KNOCK, WHO'S THERE?

To spot an AI in the wild, it's important to know the difference between **machine learning algorithms** (what we're calling AI in this book) and traditional (what programmers call **rules-based**) programs. If you've ever done basic programming, or even used HTML to design a website, you've used a rules-based program. You create a list of commands, or rules, in a language the computer can understand, and the computer does exactly what you say. To solve a problem with a rules-based program, you have to know every step required to complete the program's task and how to describe each one of those steps.

But a machine learning algorithm figures out the rules for itself via trial and error, gauging its success on goals the programmer has specified. The goal could be a list of examples to imitate, a game score to increase, or anything else. As the AI tries to reach this goal, it can discover rules and correlations that the programmer didn't even know existed. Programming an AI is almost more like teaching a child than programming a computer.

Rules-based programming

Let's say I wanted to use the more familiar rules-based programming to teach a computer to tell knock-knock jokes. The first thing I'd do is figure out all the rules. I'd analyze the structure of knock-knock jokes and discover that there's an underlying formula, as follows:

```
Knock, knock.
Who's there?
[Name]
[Name] who?
[Name] [Punchline]
```

Once I set this formula in stone, there are only two slots free that the program can control: [Name] and [Punchline]. Now the problem is reduced to just generating these two items. But I still need rules for generating them.

I could set up a list of valid names and a list of valid punchlines, as follows:

Names	Punchlines
Lettuce	in, it's cold out here!
Harry	up, it's cold out here!
Dozen	anybody want to let me in?
Orange	you going to let me in?

Now the computer can produce knock-knock jokes by choosing a name–punchline pair from the list and slotting it into the template. This doesn't create *new* knock-knock jokes but only gives me jokes I already know. I might try making things interesting by allowing [it's cold out here!] to be replaced with a few different phrases: [I'm being attacked by eels!] and [lest I transform into an unspeakable eldritch horror]. Then the program can generate a new joke:

```
Knock, knock.
Who's there?
Harry.
Harry who?
Harry up, I'm being attacked by eels!
```

I could replace [eels] with [an angry bee] or [a manta ray] or any number of things. Then I can get the computer to generate even more new jokes. With enough rules, I could potentially generate hundreds of jokes.

Depending on the level of sophistication I'm going for, I could spend a lot of time coming up with more advanced rules. I could find a list of existing puns and figure out a way to transform them into punchline format. I could even try programming in pronunciation rules, rhymes, semihomophones, cultural references, and so forth in an attempt to get the computer to recombine them into interesting puns. If I'm clever about it, I can get the program to generate new puns that nobody's ever thought of. (Although one person who tried this discovered that the algorithm's list of sayings contained words and phrases that were so old or obscure that almost nobody could understand its jokes.) No matter how sophisticated my joke-making rules get, though, I'm still telling the computer exactly how to solve the problem.

Training AI

But when I train AI to tell knock-knock jokes, I don't make the rules. The AI has to figure out those rules on its own.

The only thing I give it is a set of existing knock-knock jokes and instructions that are essentially, "Here are some jokes; try to make more of these." And the materials I give it to work with? A bucket of random letters and punctuation.

Then I leave to get coffee.

The AI gets to work.

The first thing it does is try to guess a few letters of a few knock-knock jokes. It's guessing 100 percent randomly at this point, so this first guess could be anything. Let's say it guesses something like "qasdnw,m sne?mso d." As far is it knows, this is how you tell a knock-knock joke.

Then the AI looks at what those knock-knock jokes are *actually* supposed to be. Chances are it's very wrong. "All right," says the AI, and it subtly adjusts its own structure so that it will guess a little more accurately next time. There's a limit to how drastically it can change itself, because we don't want it to try to memorize every new chunk of text it sees. But with a minimum of tweaking, the AI can discover that if it guesses nothing but k's and spaces, it will at least be right some of the time. After looking at one batch of knock-knock jokes and making one round of corrections, its idea of a knock-knock joke looks something like this:

```
    k k k  k   k
kk      k kkkok
k kkkk
k


kk
   kk   k   kk


keokk    k


   k
   k
```

Now, it's not the world's greatest knock-knock joke. But with this as a starting point, the AI can move on to a second batch of knock-knock jokes, then another. Each time, it adjusts its joke formula to improve its guesses.

After a few more rounds of guessing and self-adjusting, it has learned more rules. It has learned to employ the occasional question mark at the end of a line. It is beginning to learn to use vowels (*o* in particular). It's even making an attempt at using apostrophes.

```
noo,
Lnoc noo
Kor?
hnos h nc
pt'b oa to'

asutWtnl
toy nooc
doc kKe
```

w ce

e

How well did its rules about generating knock-knock jokes match reality? It still seems to be missing something.

If it wants to get closer to generating a passable knock-knock joke, it's going to have to figure out some rules about which *order* the letters come in. Again, it starts by guessing. The guess that *o* is always followed by *q*? Not so great, it turns out. But then it guesses that *o* is often followed by *ck*. Gold. It has made some progress. Behold its idea of the perfect joke:

```
Whock
Whock
Whock
Whock
Whock Whock Whock
Whock Whock
Whock
Whock
```

It's not quite a knock-knock joke — it sounds more like some kind of chicken. The AI's going to need to figure out some more rules.

It looks at the dataset again. It tries new ways to use "ock," looking for new combinations that match its example knock-knock jokes better.

```
nooc hock hork
aKoo kWhoc
Khock
hors Whnes Whee
noo oooc
Koo?
```

```
loo
noc no'c Wno h orea
ao no oo k KeockWnork Koos
Thoe
nock K ock
Koc
hoo
KKock Knock Thock Whonk
```

All this progress happens in just a few minutes. By the time I return with my coffee, the AI has *already* discovered that starting with "Knock Knock / Who's There?" fits the existing knock-knock jokes really, really well. It decides to apply this rule *always*.

But the rest of the formula takes quite a while to figure out, and the AI frequently devolves into playing a game of telephone with itself — or acting out a robot version of the "Who's on First?" comedy routine:

```
Knock Knock
Who's There?
Iane
Aatar who?
Aaane who?
Aan who?
Anac who?
Iobe who?
Irata who?
```

Soon it figures out the rest of the formula, but no actual, you know, *words*:

```
Knock Knock
Who's There?
```

```
Bool
Hane who?
Scheres are then the there
```

It also sometimes gets a bit carried away with the length of the punchline.

```
Knock Knock
Who's There?
Cheed
I?len who?
Ita gee the lasd emotory oud ane weonges oak anct aie
    tce suore theis onithit the where your thasg tate
    tine if thele ter the theung!
```

At some point in its training, it learns the joke about the Cow with No Lips, which becomes its Favorite Thing Ever. If you're not familiar, the joke about the Cow with No Lips goes something like:

```
Knock, knock.
Who's there?
A cow with no lips.
A cow with no lips who?
A cow with no lips says ooo ooo ooooooo.
```

According to the AI, this is comedy gold. So the AI spends a while thinking that the punchline to Every Single Joke is "ooo ooo ooooooo"...

```
Knock Knock
Who's There?
Alec
Anic who?
```

Wow co poo the tho tho door to the soout oo ooo oooo

   ooo oo oo oo oo ooo oo oo oo ooo ooo oooo ooo ooo oo

   oo ooo ooo ooo ooo ooo oo oo ooo ooo ooo ooo oo oo

   oo ooo ooo ooo ooo . . .

It has discovered a rule, all right. Eventually it (reluctantly, I assume) tones down how often it uses the Cow with No Lips rule to make punchlines. But it still has problems that are remarkably similar to two people telling knock-knock jokes over a very bad phone connection, maybe forgetting who was actually telling the joke:

Knock Knock
Who's There?
Alther
Water who?
Water who who?
Who who?
Arton and out on the moor?

At last it figures out the structure of a knock-knock joke and proceeds to make jokes that, while technically correct, don't make all that much sense. Many of them are partially plagiarized from jokes in the training dataset.

Knock Knock
Who's There?
Robin
Robin who?
Robin you! Git me and I'm leaving

Knock Knock
Who's There?

```
Bet
Beef who?
Beef ano with no lips aslly.

Ireland
Ireland who?
Ireland you money, butt.
```

And then. It produced. An actual joke. That it had composed entirely on its own, without plagiarizing from the dataset, and that was not only intelligible but also actually…funny?

```
Knock Knock
Who's There?
Alec
Alec who?
Alec- Knock Knock jokes.
```

Did the AI suddenly begin to understand knock-knock jokes and English-language puns? Almost definitely not, given the very small size of the dataset. But the freedom that the AI had — free rein over the entire set of possible characters — allowed it to try new combinations of sounds, one of which ended up actually working. So more of a victory for the infinite monkey theory* than a proof of concept for the next AI-only comedy club.

---

* The old adage that a monkey writing randomly on a typewriter for an infinite amount of time will eventually produce the entire works of Shakespeare actually pretty accurately describes the "brute force" method of searching for a solution to a problem by systematically trying everything. Ideally, using AI to solve the problem is an improvement over this. Ideally.

The beauty of letting AI make its own rules is that a single approach—here's the data; you try to figure out how to copy it—works on a lot of different problems. If I had given the joke-telling algorithm another dataset instead of knock-knock jokes, it would have learned to copy that dataset instead.

It could create new species of birds:

```
Yucatan Jungle-Duck
Boat-billed Sunbird
Western Prong-billed Flowerpecker
Black-capped Flufftail
Iceland Reedhaunter
Snowy Mourning Heron-Robin
```

Or new perfumes:

```
Fancy Ten
Eau de Boffe
Frogant Flower
Momite
Santa for Women
```

Or even new recipes.

BASIC CLAM FROSTING
```
main dish, soups
1 lb chicken
1 lb pork, cubed
½ clove garlic, crushed
1 cup celery, sliced
1 head (about ½ cup)
```

6 tablespoon electric mixer

1 teaspoon black pepper

1 onion—chopped

3 cup beef broth the owinls for a fruit

1 freshly crushed half and half; worth water

With pureed lemon juice and lemon slices in a 3-quart pan.

Add vegetables, add chicken to sauce, mixing well in onion. Add bay leaf, red pepper, and slowly cover and simmer covered for 3 hours. Add potatoes and carrots to simmering. Heat until sauce boils. Serve with pies.

It the liced pieces cooked up desserts, and cook over wok.

Refrigerate up to ½ hour decorated.

Yield: 6 servings

JUST LET THE AI FIGURE IT OUT

Given a set of knock-knock jokes and no further instruction, the AI was able to discover a lot of the rules that I would have otherwise had to manually program into it. Some of its rules I would never have thought to program in or wouldn't even have known existed — such as The Cow with No Lips Is the Best Joke.

This is part of what makes AIs attractive problem solvers, and is particularly handy if the rules are really complicated or just plain mysterious. For example, AI is often used for image recognition, a surprisingly complicated task that's difficult to do with an ordinary computer program. Although most of us are easily able to identify a cat in a picture, it's really hard to come up with the rules that define a cat. Do we tell the program that a cat has two eyes, a nose, two ears, and a tail? That also describes a mouse and a giraffe. And what if the cat is curled up or facing away? Even writing down the rules for detecting a single eye is tricky. But an AI can look at tens of thousands of images of cats and come up with rules that correctly identify a cat most of the time.

Sometimes AI is only a small part of a program while the rest of it is rules-based scripting. Consider a program that lets customers call their banks for account information. The voice-recognition AI matches spoken sounds to options in the help-line menu, but programmer-issued rules govern the list of options the caller can access and the code that identifies the account as belonging to the customer.

Other programs start out as AI-powered but switch control over to humans if things get tough, an approach called pseudo-AI. Some customer-service chat windows work like this. When you begin a conversation with a bot, if you act too confusing, or if the AI detects that you are getting annoyed, you may suddenly find yourself chatting with a human instead. (A human who unfortunately now has to deal with a confused and/or annoyed customer—maybe a "talk to a human" option would be better for customer *and* employee.) Today's self-driving cars work this way, too—the driver has to always be ready to take control if the AI gets flustered.

AI is also great at strategy games like chess, for which we know how to describe all possible moves but not how to write a formula that tells us what the best next move is. In chess, the sheer number of possible moves and complexity of game play means that even a grandmaster would be unable to come up with hard-and-fast rules governing the best move in any given situation. But an algorithm can play a bunch of practice games against itself — millions of them, more than even the most dedicated grandmaster — to come up with rules that help it win. And since the AI learned without explicit instruction, sometimes its strategies are very unconventional. Sometimes a little *too* unconventional.

If you don't tell AI which moves are valid, it may find and exploit strange loopholes that completely break your game. For example, in 1997 some programmers built algorithms that could play tic-tac-toe remotely against each other on an infinitely large board. One programmer, rather than designing a rules-based strategy, built an AI that could evolve its own approach. Surprisingly, the AI suddenly began winning all its games. It turned out that the AI's strategy was to place its move very, very far away, so that when its

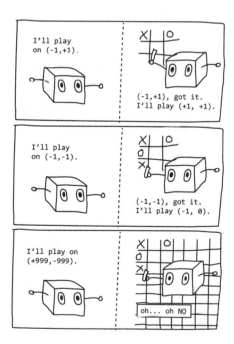

opponent's computer tried to simulate the new, greatly expanded board, the effort would cause it to run out of memory and crash, forfeiting the game.[1] Most AI programmers have stories like this—times when their algorithms surprised them by coming up with solutions they hadn't expected. Sometimes these new solutions are ingenious, and sometimes they're a problem.

At its most basic, all AI needs is a goal and a set of data to learn from and it's off to the races, whether the goal is to copy examples of loan decisions a human made or predict whether a customer will buy a certain sock or maximize the score in a video game or maximize the distance a robot can travel. In every scenario, AI uses trial and error to invent rules that will help it reach its goal.

SOMETIMES ITS RULES ARE BAD

Sometimes, an AI's brilliant problem-solving rules actually rely on mistaken assumptions. For example, some of my weirdest AI experiments have involved Microsoft's image recognition product, which allows you to submit any image for the AI to tag and caption. Generally, this algorithm gets things right—identifying clouds, subway trains, and even a kid doing some sweet skateboarding tricks. But one day I noticed something odd about its results: it was tagging sheep in pictures that definitely did not contain any sheep. When I investigated further, I discovered that it tended to see sheep in landscapes that had lush green fields—whether or not the sheep were actually there. Why the persistent—and specific—error? Maybe during training this AI had mostly been shown sheep that were in fields of this sort and had failed to realize that the "sheep" caption referred to the animals, not to the grassy landscape. In other words, the AI had been looking at the wrong thing. And sure enough, when I showed it examples of sheep that were *not* in lush green fields, it tended to get confused. If I showed it pictures of sheep in cars, it would tend to label them as dogs or cats instead. Sheep in living rooms also got labeled as dogs and cats, as did sheep held in people's arms. And sheep on leashes were identified as dogs. The AI also had similar problems with goats — when they climbed into trees, as they sometimes do, the algorithm thought they were giraffes (and another similar algorithm called them birds).

A herd of sheep grazing on a lush green landscape

A herd of sheep grazing on a lush green landscape

Although I couldn't know for sure, I could guess that the AI had come up with rules like Green Grass = Sheep, and Fur in Cars or Kitchens = Cats. These rules had served it well in training but failed when it encountered the real world and its dizzying variety of sheep-related situations.

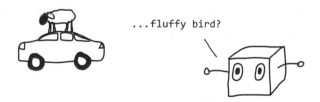

...fluffy bird?

Training errors like these are common with image recognition AIs. But the consequences of these mistakes can be serious. A team at Stanford University once trained an AI to tell the difference between pictures of healthy skin and pictures of skin cancer. After the researchers trained their AI, however, they discovered that they had inadvertently trained a ruler detector instead — many of the tumors in their training data had been photographed next to rulers for scale.[2]

HOW TO DETECT A BAD RULE

It's often not that easy to tell when AIs make mistakes. Since we don't write their rules, they come up with their own, and they don't write them down or explain them the way a human would. Instead, the AIs make complex interdependent adjustments to their own internal structures, turning a generic framework into something that's fine-tuned for an individual task. It's like starting with a kitchen full of generic ingredients and ending with cookies. The rules might be stored in the connections between virtual brain cells or in the genes of a virtual organism. The rules might be complex, spread out, and weirdly entangled with one another. Studying an AI's internal structure can be a lot like studying the brain or an ecosystem — and you don't need to be a neuroscientist or an ecologist to know how complex those can be.

Researchers are working on finding out just how AIs make decisions, but in general, it's hard to discover what an AI's internal rules actually are. Often it's merely because the rules are hard to understand, but at other times, especially when it comes to commercial and/or government algorithms, it's because the algorithm itself is proprietary. So unfortunately, problems often turn up in the algorithm's results when it's already in use, sometimes making decisions that can affect lives and potentially cause real harm.

For example, an AI that was being used to recommend which prisoners would be paroled was found to be making prejudiced decisions, unknowingly copying the racist behaviors it found in its training.[3] Even without understanding what bias is, AI can still manage to be biased. After all, many AIs learn by copying humans. The question they're answering is not "What is the best solution?" but "What would the humans have done?"

Systematically testing for bias can help catch some of these common problems before they do damage. But another piece of the puzzle is learning to anticipate problems before they occur and designing AIs to avoid them.

FOUR SIGNS OF AI DOOM

When people think of AI disaster, they think of AIs refusing orders, deciding that their best interests lie in killing all humans, or creating terminator bots. But all those disaster scenarios assume a level of critical thinking and a humanlike understanding of the world that AIs won't be capable of for the foreseeable future. As leading machine learning researcher Andrew Ng put it, worrying about an AI takeover is like worrying about overcrowding on Mars.[4]

That's not to say that today's AIs can't cause problems. From slightly annoying their programmers all the way to perpetuating prejudices or crashing a self-driving car, today's AIs are not exactly harmless. But by knowing a little about AI, we can see some of these problems coming.

Here's how an AI disaster might actually play out today.

Let's say a Silicon Valley startup is offering to save companies time by screening job candidates, identifying the likely top performers by analyzing short video interviews. This could be attractive — companies spend a lot of time and resources interviewing dozens of candidates just to find that one good match. Software never gets tired and never gets hangry, and it doesn't hold personal grudges. But what are the warning signs that what the company is building is actually an AI disaster?

Warning sign number 1: The Problem Is Too Hard

The thing about hiring good people is that it's really difficult. Even humans have trouble identifying good candidates. Is this candidate genuinely excited to work here or just a good actor? Have we accounted for disability or differences in culture? When you add AI to the mix, it gets even more difficult. It's nearly impossible for AI to understand the nuances of jokes or tone or cultural references. And what if a candidate makes a reference to the day's current events? If the AI was trained on data collected the previous year, it won't have a chance of understanding — and it might punish the candidate for saying something it finds nonsensical. To do the job well, the AI will have to have a huge range of skills and keep track of a large amount of information. If it isn't capable of doing the job well, we're in for some kind of failure.

Warning sign number 2: The Problem Is Not What We Thought It Was

The problem with designing an AI to screen candidates for us: we aren't really asking the AI to identify the best candidates. We're asking it to identify the candidates that most resemble the ones our human hiring managers liked in the past.

That might be okay if the human hiring managers made great decisions. But most US companies have a diversity problem, particularly among managers and particularly in the way that hiring managers evaluate resumes and interview candidates. All else being equal, resumes with white-male-sounding names are more likely to get interviews than those with female-and/or minority-sounding names.[5] Even hiring managers who are female and/or members of a minority themselves tend to unconsciously favor white male candidates.

Plenty of bad and/or outright harmful AI programs are designed by people who thought they were designing an AI to solve a problem but were unknowingly training it to do something entirely different.

Warning sign number 3: There Are Sneaky Shortcuts

Remember the skin-cancer-detecting AI that was really a ruler detector? Identifying the minute differences between healthy cells and cancer cells is difficult, so the AI found it a lot easier to look for the presence of a ruler in the picture.

If you give a job-candidate-screening AI biased data to learn from (which you almost certainly did, unless you did a lot of work to scrub bias from the data), then you also give it a convenient shortcut to improve its accuracy at predicting the "best" candidate: prefer white men. That's a lot easier than analyzing the nuances of a candidate's choice of wording. Or perhaps the AI will find and exploit another unfortunate shortcut — maybe we filmed our successful candidates using a single camera, and the AI learns to read the camera metadata and select only candidates who were filmed with that camera.

AIs take sneaky shortcuts all the time — they just don't know any better!

Warning sign number 4: The AI Tried to Learn from Flawed Data

There's an old computer-science saying: garbage in, garbage out. If the AI's goal is to imitate humans who make flawed decisions, perfect success would be to imitate those decisions exactly, flaws and all.

Flawed data, whether it's flawed examples to learn from or a flawed simulation with weird physics, will throw an AI for a loop or send it off in the wrong direction. Since in many cases our example data *is* the problem we're giving the AI to solve, it's no wonder that bad data leads to a bad solution. In fact, warning signs numbers 1 through 3 are most often evidence of problems with data.

DOOM — OR DELIGHT

The job-candidate-screening example is, unfortunately, not hypothetical. Multiple companies already offer AI-powered resume-screening or video-interview-screening services, and few offer information about what they've done to address bias or to account for disability or cultural differences or to find out what information their AIs use in the screening process. With careful work, it's at least possible to build a job-candidate-screening AI that is measurably less biased than human hiring managers — but without published stats to prove it, we can be pretty sure that bias is still there.

The difference between successful AI problem solving and failure usually has a lot to do with the suitability of the task for an AI solution. And there are plenty of tasks for which AI solutions are more efficient than human solutions. What are they, and what makes AI so good at them? Let's take a look.

CHAPTER 2

AI Is everywhere, but where Is It exactly?

THIS EXAMPLE IS REAL, I KID YOU NOT

There's a farm in Xichang, China, that's unusual for a number of reasons. One, it's the largest farm of its type in the world, its productivity unmatched. Each year, the farm produces six billion *Periplaneta americana*, more than twenty-eight thousand of them per square foot.[1] To maximize productivity, the farm relies on algorithms that control the temperature, humidity, food supply, and even analyze the genetics and growing rate of *Periplaneta americana*.

But the primary reason the farm is unusual is that *Periplaneta americana* is simply the Latin name for the common cockroach. Yes, the farm produces cockroaches, which are crushed into a potion that's highly valuable in traditional Chinese medicine. "Slightly sweet," reports its packaging. With "a slightly fishy smell."

Because it's a valuable trade secret, details are scarce on what exactly the cockroach-maximizing algorithm is like. But the scenario sounds an awful lot like a famous thought experiment called the paper-clip maximizer, which supposes that a superintelligent AI has a singular task: producing paper clips. Given that single-minded goal, a superintelligent AI *might* decide to convert all the resources it could into the manufacture of paper clips — even converting the planet and all its occupants into paper clips. Fortunately — *very* fortunately, given that we've just been talking about an algorithm whose job it is to maximize the number of cockroaches in existence — the algorithms we have today are light-years away from being capable of running factories or farms by themselves, let alone converting the global economy into a cockroach producer. Very likely, the cockroach AI is making predictions about future production rates based on past data, then picking the environmental conditions it thinks will maximize cockroach production. It likely can suggest adjustments within a range that its human engineers set, but it probably relies on humans for taking data, filling orders, unloading supplies, and the all-important marketing of cockroach extract.

Revenge will be ours.

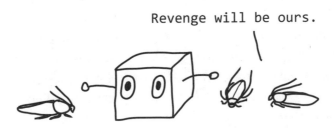

Still, helping optimize a cockroach farm is something an AI is likely to be good at. There's a lot of data to parse, but these algorithms are good at finding trends in huge datasets. It's a job that is likely to be unpopular, but

AIs don't mind repetitive tasks or the skittering sound of millions of cockroach feet in the dark. Cockroaches reproduce quickly, so it doesn't take long to see the effects of variable tweaking. And it's a specific, narrow problem rather than one that's complex and open-ended.

Are there still potential problems with using AI to maximize cockroach production? Yes. Since AIs lack context about what they're actually trying to accomplish and why, they often solve problems in unexpected ways. Suppose the cockroach AI found that by turning both the heat and water up to "max" in one particular room, it can significantly increase the number of cockroaches that room can produce. It would have no way of knowing (or caring) that what it had actually done was short out the door that prevents the cockroaches from accessing the employee kitchen.

Technically, shorting out the door was the AI being good at its job. Its job was to maximize cockroach production, not guard against their escape. To work with AI effectively, and to anticipate trouble before it happens, we need to understand what machine learning is best at.

ACTUALLY, I WOULD BE FINE WITH A ROBOT TAKING *THIS* JOB

Machine learning algorithms are useful even for jobs that a human could do better. Using an algorithm for a particular task saves the trouble and expense of having a human do it, especially when the task is high-volume and repetitive. This is true not just for machine learning algorithms, of course, but for automation in general. If a Roomba can save us from having to vacuum a room ourselves, we'll put up with retrieving it again and again from under the sofa.

One repetitive task that people are automating with AI is analyzing medical images. Lab technicians spend hours every day looking at blood samples under a microscope, counting platelets or white or red blood cells or examining tissue samples for abnormal cells. Each one of these tasks is simple, consistent, and self-contained, so in that way they're good candidates for automation. But the stakes are higher when these algorithms leave the research lab and start working in hospitals, where the consequences of a mistake are much more serious. There are similar problems with self-driving cars — driving is mostly repetitive, and it would be nice to have a driver who never gets tired, but even a tiny glitch can have serious consequences at sixty miles per hour.

Another high-volume task we're happy to automate with AI, even if its performance isn't quite at the human level: spam filtering. The onslaught of spam is a problem that can be nuanced and ever-changing, so it's a tricky one for AI, but on the other hand, most of us are willing to put up with the occasional misfiltered message if it means our inboxes are mostly clear. Flagging malicious URLs, filtering social media posts, and identifying bots are high-volume tasks in which we mostly tolerate buggy performance.

Hyperpersonalization is another area where AI is starting to show its usefulness. With personalized product recommendations, movie recommendations, and music playlists, companies use AI to tailor the experience to each consumer in a way that would be cost-prohibitive if a human were coming up with the requisite insights. So what if the AI is convinced that we need an endless number of hallway rugs or thinks we are a toddler because of that one time we bought a present for a baby shower? Its mistakes are mostly harmless (except for those occasions when they're very, very unfortunate), and it could bring the company a sale.

Perhaps you'd like this book
that's just like one you
bought and hated?

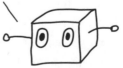

Commercial algorithms can now write up hyperlocal articles about election results, sports scores, and recent home sales. In each case, the algorithm can only produce a highly formulaic article, but people are interested enough in the content that it doesn't seem to matter. One of these algorithms is called Heliograf, developed by the *Washington Post* to turn sports stats into news articles. As early as 2016, it was already producing hundreds of articles a year. Here's an example of its reporting on a football game.[2]

The Quince Orchard Cougars shut out the Einstein Titans, 47–0, on Friday.

Quince Orchard opened the game with an eight-yard touchdown off a blocked punt return by Aaron Green. The Cougars added to their lead on Marquez Cooper's three-yard touchdown run. The Cougars extended their lead on Aaron Derwin's 18-yard touchdown run. The Cougars went even further ahead following Derwin's 63-yard touchdown reception from quarterback Doc Bonner, bringing the score to 27–0.

It's not exciting stuff, but Heliograf does describe the game.* It knows how to populate an article based on a spreadsheet full of data and a few stock sports phrases. But an AI like Heliograf would utterly fail when faced with information that doesn't fit neatly into the prescribed boxes. Did a horse run onto the field midgame? Was the locker room of the Einstein Titans overrun by cockroaches? Is there an opportunity for a clever pun? Heliograf only knows how to report its spreadsheet.

Nevertheless, AI-generated writing allows news outlets to produce the

---

* The fact that the score is 27–0 at this point rather than 28–0 means that the Cougars might have missed one of their conversion points — a fact that Heliograf fails to mention.

types of articles that were formerly cost-prohibitive. It requires a human's touch to decide which articles to automate and to build the AI's basic templates and stock phrases, but once a paper has set up one of these hyperspecialized algorithms, it can churn out as many news articles as there are spreadsheets to draw from. One Swedish news site, for example, built the Homeowners Bot, which was able to read tables of real estate data and write up each sale into an individual article, producing more than ten thousand articles in four months. This has turned out to be the most popular — and lucrative — type of article the news site publishes.[3] And human reporters can spend their valuable time on creative investigative work instead. Increasingly, major news outlets use AI assistance to write their articles.[4]

Science is another area where AI shows promise for automating repetitive tasks. Physicists, for example, have used AI to watch the light coming from distant stars,[5] looking for telltale signs that the star might have a planet. Of course, the AI wasn't as accurate as the physicists who trained it. Most of the stars it flagged as interesting were false alarms. But it was able to correctly eliminate more than 90 percent of the stars as *un*interesting, which saved the physicists a lot of time.

Astronomy is full of huge datasets, as it turns out. Over the course of its life, the Euclid telescope will collect tens of billions of galaxy images, out of which maybe two hundred thousand will show evidence of a phenomenon called gravitational lensing,[6] which happens when a supermassive galaxy has gravity so strong that it actually bends the light from other, more distant galaxies. If astronomers can find the lenses, they can learn a lot about gravity on a huge intergalactic scale, where there are so many unsolved mysteries that a full 95 percent of the universe's mass and energy is unaccounted for. When algorithms reviewed the images, they were faster than humans and sometimes outperformed them in accuracy. But when the telescope captured one superexciting "jackpot" lens, only the humans noticed it.

Creative work can be automated as well, at least under the supervision of a human artist. Whereas before a photographer might spend hours tweaking a photograph, today's AI-powered filters, like the built-in ones on Instagram and Facebook, do a decent job of adjusting contrast and lighting and even adding depth-of-focus effects to simulate an expensive lens. No need to digitally paint cat ears onto your friend — there's an AI-powered filter built into your Instagram that will figure out where the ears should go, even as your friend moves their head. In big and small ways, AI gives artists and musicians access to time-saving tools that can expand their ability to do creative work on their own. On the flip side of this, of course, are tools like **deepfakes,** which allow people to swap one person's head and/or body for another, even in video. On the one hand, greater access to this tool means that artists can readily insert Nicolas Cage or John Cho into various movie roles, goofing around or making a serious point about minority representation in Hollywood.[7] On the other hand, the increasing ease of deepfakes is already giving harassers new ways to generate disturbing, highly targeted videos for dissemination online. And as technology improves and deepfake videos become increasingly convincing, many people and governments are worrying about the technique's potential for creating fake but damaging videos — like realistic yet faked videos of a politician saying something inflammatory.

In addition to saving humans time, AI automation can mean more consistent performance. After all, an individual human's performance may vary throughout the day depending on things like how recently they've eaten or how much they've slept, and each person's biases and moods might have a huge effect as well. Countless studies have shown that sexism, racial bias, ableism, and other problems affect things like whether resumes get shortlisted, whether employees get raises, and whether prisoners get parole. Algorithms avoid human inconsistencies — given a set of data, they'll return pretty much an unvarying result, no matter if it's morning, noon, or happy hour. But, unfortunately, consistent doesn't mean unbiased. It's very possible for an algorithm to be consistently unfair, especially if it learned, as many AIs do, by copying humans.

So there are plenty of things that it's attractive to automate with AI. But what about the things that determine whether we *can* automate a problem?

THE NARROWER THE TASK, THE SMARTER THE AI

The **Turing test** has been a famous benchmark for the intelligence level of a computer program ever since Alan Turing proposed it in the 1950s. A computer program passes the standard Turing test if it can chat with humans and convince approximately one-third of them that it's a human being rather than a computer. Passing the Turing test is sometimes seen as a sign that an algorithm has achieved human-level intelligence and maybe even self-awareness. Lots of science fiction books and movies — *Blade Runner, Ex Machina, Bicentennial Man,* and many more — involve sophisticated artificial general intelligences that have proved their "personhood" by passing the Turing test.

But the Turing test isn't actually a good measure of algorithmic intelligence. For one thing, it's easy to pass the Turing test if you can make the topic of conversation narrow enough. I chatted with the Whole Foods Market bot on Facebook to test this theory:

> **Whole Foods:** Hi Janelle! We're here to help you find recipes as simple as their ingredients.
> **Me:** Do you have a recipe for guacamole?
> **Whole Foods:** <posts a recipe: Guacamole with Cilantro>
> **Me:** Is it okay to make guacamole from green peas?
> **Whole Foods:** <posts a recipe: Green Pea Guacamole>

Me: Give me a recipe that uses avocado. Anything but guacamole.

Whole Foods: <posts a recipe: Avocado Salsa with Olives and Cilantro>

At this point, the conversation is a bit ambiguous. The thing I'm chatting with hasn't specifically said whether it's a human or a bot and is competently handling questions that can be answered by posting a recipe. But when I stray from this narrow format, the bot quickly reveals itself.

Me: How can I tell if an avocado is ripe?

Whole Foods: <posts a recipe: Edamame Guacamole>

Me: Do you have a favorite Star Wars character?

Whole Foods: <posts a recipe: Sautéed Shrimp with Polenta and Manchego>

I *loved* Sautéed Shrimp with Polenta and Manchego in The Last Jedi!

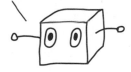

This is the strategy companies use when they want to use chatbots for customer service. Rather than identify the bots as such, they rely on human politeness to keep the conversation on topics in which the bots can hold their own. After all, if there's a chance you might be talking with a human employee, it would be rude to test them with weird off-topic questions.

Even when customers stick to the prescribed topic, chatbots will struggle if the topic is too broad. Beginning in August 2015, Facebook tried to create an AI-powered chatbot called M that was meant to make hotel reservations, book theater tickets, recommend restaurants, and more.[8] The idea was that the company would start out using humans to handle the most difficult requests, thereby generating lots of examples that the algorithm could learn from. Eventually, Facebook expected the algorithm to have enough data to handle most questions on its own. Unfortunately, given the freedom to ask M anything, customers took Facebook at its word. In an interview, the engineer who started the project recounted, "People try first to ask for the weather tomorrow; then they say 'Is there an Italian restaurant available?' Next they have a question about immigration, and after a while they ask M to organize their wedding."[9] A user even asked M to arrange for a parrot to visit his friend. M succeeded — by sending that request to be handled by a human. In fact, years after it introduced M, Facebook found that its algorithm still needed too much human help. It shut down the service in January 2018.[10]

um this is
not a parrot

Dealing with the full range of things a human can say or ask is a very broad task. The mental capacity of AI is still tiny compared to that of humans, and as tasks become broad, AIs begin to struggle.

For example, I recently trained an AI to generate recipes. This particular AI is set up to imitate text, but it started from a blank slate — no idea what recipes are, no idea that various letters are referring to ingredients and things that happen to them, no idea even what English is. There's a lot to keep track of, but it tried its best to figure out how to place one letter after another and imitate the recipes it saw. When I gave it only recipes for cake to learn from, here's the recipe it produced.

Carrot Cake (Vera Ladies"
cakes, alcohol
1 pkg yellow cake mix
3 cup flour
1 teaspoon baking powder
1 ½ teaspoon baking soda
¼ teaspoon salt
1 teaspoon ground cinnamon
1 teaspoon ground ginger
½ teaspoon ground cloves
1 teaspoon baking powder
½ teaspoon salt
1 teaspoon vanilla
1 egg, room temperature
1 cup sugar
1 teaspoon vanilla
1 cup chopped pecans

Preheat oven to 350 degrees. Grease a 9-inch springform pan.

To make the cake: Beat eggs at high speed until thick and yellow color and set aside. In a separate bowl, beat the egg whites until stiff. Speed the first like the mixture into the prepared pan and smooth the batter. Bake in the oven for about 40 minutes or until a wooden toothpick inserted into centre comes

out clean. Cool in the pan for 10 minutes.
Turn out onto a wire rack to cool completely.

Remove the cake from the pan to cool
completely. Serve warm.

HereCto Cookbook (1989) From the Kitchen &
Hawn inthe Canadian Living

Yield: 16 servings

Now, the recipe isn't perfect, but at least it's a recipe that's identifiably cake (even if, when you look at the instructions closely, you realize that it only produces a single baked egg yolk).

Next, I asked the AI to learn to generate not just cake recipes but also recipes for soup, barbecue, cookies, and salads. It had about ten times more data to learn from — 24,043 general recipes as opposed to just 2,431 recipes in the cake-only dataset. Here's a recipe it generated.

Spread Chicken Rice
cheese/eggs, salads, cheese
2 lb hearts, seeded
1 cup shredded fresh mint or raspberry pie
½ cup catrimas, grated
1 tablespoon vegetable oil
1 salt
1 pepper
2 ½ tb sugar, sugar

Combine unleaves, and stir until the mixture
is thick. Then add eggs, sugar, honey, and
caraway seeds, and cook over low heat. Add
the corn syrup, oregano, and rosemary and

the white pepper. Put in the cream by heat.
Cook add the remaining 1 teaspoon baking
powder and salt. Bake at 350F for 2 to 1 hour.
Serve hot.

Yield: 6 servings

This time, the recipe is a total disaster. The AI had to try to figure out when to use chocolate and when to use potatoes. Some recipes required baking, some required slow simmering, and the salads required no cooking at all. With all these rules to try to learn and keep track of, the AI spread its brainpower too thin.

So people who train AIs to solve commercial or research problems have discovered that it makes sense to train it to specialize. If an algorithm seems to be better at its job than the AI that invented Spread Chicken Rice, the main difference is probably that it has a narrower, better-chosen problem. The narrower the AI, the smarter it seems.

C-3PO VERSUS YOUR TOASTER

This is why AI researchers like to draw a distinction between **artificial narrow intelligence (ANI),** the kind we have now, and **artificial general intelligence (AGI),** the kind we usually find in books and movies. We're used to stories about superintelligent computer systems like Skynet and Hal or very human robots like Wall-E, C-3PO, Data, and so forth. The AIs in these stories may struggle to understand the fine points of human emotion, but they're able to understand and react to a huge range of objects and situations. An AGI could beat you at chess, tell you a story, bake you a cake, describe a sheep, and name three things larger than a lobster. It's also solidly the stuff of science fiction, and most experts agree that AGI is many decades away from becoming reality — if it will become a reality at all.

The ANI that we have today is less sophisticated. Considerably less sophisticated. Compared to C-3PO, it's basically a toaster.

The algorithms that make headlines when they beat people at games like chess and go, for example, surpass humans' ability at a single specialized task. But machines have been superior to humans at specific tasks for a while now. A calculator has always exceeded humans' ability to perform long division — but it still can't walk down a flight of stairs.

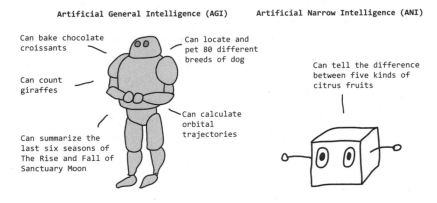

Artificial General Intelligence (AGI)   Artificial Narrow Intelligence (ANI)

Can bake chocolate croissants

Can count giraffes

Can summarize the last six seasons of The Rise and Fall of Sanctuary Moon

Can locate and pet 80 different breeds of dog

Can calculate orbital trajectories

Can tell the difference between five kinds of citrus fruits

Actually, plenty of sci-fi AGIs are for some reason unable to walk down stairs. The Daleks, C-3PO, *RoboCop* thingy, Hal. Study further?

What problems are narrow enough to be suitable for today's ANI algorithms? Unfortunately (see warning sign number 1 of AI doom: Problem Is Too Hard), often a real-world problem is broader than it first appears. In our video-interview-analyzing AI from chapter 1, the problem at first glance seems relatively narrow: a simple matter of detecting emotion in human faces. But what about applicants who have had a stroke or who have facial scarring or who don't emote in neurotypical ways? A human could understand an applicant's situation and adjust their expectations accordingly, but to do the same, an AI would have to know what words the applicant is saying (speech-to-text is an entire AI problem in itself), understand what those words mean (current AIs can only interpret the meaning of limited kinds of sentences in limited subject areas and don't do well with nuance), and use that knowledge and understanding to alter how it interprets emotional data. Today's AIs, incapable of such a complicated task, would most likely screen all these people out before they got to a human.

As we'll see below, self-driving cars may be another example of a problem that is broader than it at first appears.

INSUFFICIENT DATA DOES NOT COMPUTE

AIs are slow learners. If you showed a human a picture of some new animal called a wug, then gave them a big batch of pictures and told them to identify all the pictures that contain wugs, they could probably do a decent job just based on that one picture. An AI, however, might need thousands or hundreds of thousands of wug pictures before it could even semireliably identify wugs. And the wug pictures need to be varied enough for the algorithm to figure out that "wug" refers to an animal, not to the checkered floor it's standing on or to the human hand patting its head.

Researchers are working on designing AIs that can master a topic with fewer examples (an ability called **one-shot learning**), but for now, if you want to solve a problem with AI, you'll need tons and tons of training data. The popular ImageNet set of training data for image generation or image recognition currently has 14,197,122 images in only one thousand different categories. Similarly, while a human driver may only need to accumulate a few hundred hours of driving experience before they're allowed to drive on their own, as of 2018 the self-driving car company Waymo's cars have collected data from driving more than six million road miles plus five billion more miles driven in simulation.[11] And we're still a ways off from a widespread rollout of self-driving car technology. AI's data hungriness is a big reason why the age of "big data," where people collect and analyze huge sets of data, goes hand in hand with the age of AI.

Sometimes AIs learn so slowly that it's impractical to let them do their learning in real time. Instead, they learn in sped-up time, amassing hundreds of years' worth of training in just a few hours. A program called OpenAI Five, which learned to play the computer game Dota (an online fantasy game in which teams have to work together to take over a map), was able to beat some of the world's best human players by playing games against itself rather than against humans. It challenged itself to tens of thousands of simultaneous games, accumulating 180 years of gaming time each day.[12] Even if the goal is to do something in the real world, it can make sense to build a simulation of that task to save time and effort.

Another AI's task was to learn to balance a bicycle. It was a bit of a slow learner, though. The programmers kept track of all the paths the bicycle's front wheel took as it repeatedly wobbled and crashed. It took more than a hundred crashes before the AI could drive more than a few meters without falling, and thousands more before it could go more than a few tens of meters.

Training an AI in simulation is convenient, but it also comes with risks. Because of the limited computing power of the computers that run them, simulations aren't nearly as detailed as the real world and are by necessity held together with all sorts of hacks and shortcuts. That can sometimes be a problem if the AI notices the shortcuts and begins to exploit them (more on that later).

PIGGYBACKING ON OTHER PROGRESS

If you don't have lots of training data, you might still be able to solve your problem with AI if you or someone else has already solved a similar problem. If the AI starts not from scratch but from a configuration it learned from a previous dataset, it can reuse a lot of what it learned. For example, say I already have an AI that I've trained to generate the names of metal bands. If my next task is to build an AI that can generate ice cream flavors, I may get results more quickly, and need fewer examples, if I start with the metal-band AI. After all, from learning to generate metal bands, the AI already knows

- approximately how long each name should be,
- that it should capitalize the first letter of each line,
- common letter combinations — *ch* and *va* and *str* and *pis* (it is already partway to spelling *chocolate, vanilla, strawberry,* and *pistachio*!) — and
- commonly occurring words, such as *the* and, um…*death*?

So a few short rounds of training can retrain the AI from a model that produces this:

```
Dragonred of Blood
Stäggabash
Deathcrack
Stormgarden
Vermit
Swiil
Inbumblious
Inhuman Sand
Dragonsulla and Steelgosh
Chaosrug
Sespessstion Sanicilevus
```

into a model that produces this:

```
Lemon-Oreo
Strawberry Churro
Cherry Chai
Malted Black Madnesss
Pumpkin Pomegranate Chocolate Bar
Smoked Cocoa Nibe
Toasted Basil
Mountain Fig n Strawberry Twist
Chocolate Chocolate Chocolate Chocolate Road
Chocolate Peanut Chocolate Chocolate Chocolate
```

(There's only a *miiinor* awkward phase in between, when it's generating things like this:)

Swirl of Hell

Person Cream

Nightham Toffee

Feethberrardern's Death

Necrostar with Chocolate Person

Dirge of Fudge

Beast Cream

End All

Death Cheese

Blood Pecan

Silence of Coconut

The Butterfire

Spider and Sorbeast

Blackberry Burn

Maybe I should have started with pie instead.

Chocolate
Peanut
Chocolate
Chocolate
Chocolate

Beet
Bourbon

Praline
Cheddar
Swirl

As it turns out, AI models get reused a lot, a process called **transfer learning**. Not only can you get away with using less data by starting with an AI that's already partway to its goal, you can also save a lot of time. It can take days or even weeks to train the most complex algorithms with the largest datasets, even on very powerful computers. But it takes only minutes or seconds to use transfer learning to train the same AI to do a similar task.

People use transfer learning a lot in image recognition in particular, since training a new image recognition algorithm from scratch requires a lot of time and a lot of data. Often people will start with an algorithm that's already been trained to recognize general sorts of objects in generic images, then use that algorithm as a starting point for specialized object recognition. For example, if an algorithm already knows rules that help it recognize pictures of trucks, cats, and footballs, it already has a head start on the task of distinguishing different kinds of produce for, say, a grocery scanner. A lot of the rules a generic image recognition algorithm has to discover—rules that help it find edges, identify shapes, and classify textures—will be helpful for the grocery scanner.

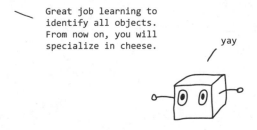

Great job learning to identify all objects. From now on, you will specialize in cheese.

yay

DON'T ASK IT TO REMEMBER

A problem is more easily solvable with AI if it doesn't require much memory. Because of their limited brainpower, AIs are particularly bad at remembering things. This shows up, for example, when AIs try to play computer games. They tend to be extravagant with their characters' lives and other resources (like powerful attacks that they only have in limited numbers). They'll burn through lots of lives and spells at first until their numbers get critically low, at which point they'll suddenly start being cautious.[13]

One AI learned to play the game Karate Kid, but it always squandered all its powerful Crane Kick moves at the beginning of the game. Why? It only had enough memory to look forward to the next six seconds of game play. As Tom Murphy, who trained the algorithm, put it, "Anything that you are gonna need 6 seconds later, well, too bad. Wasting lives and other resources is a common failure mode."[14]

Even sophisticated algorithms like OpenAI's Dota-playing bot have only a limited time frame over which they can remember and predict. OpenAI Five can predict an impressive two minutes into the future (impressive for a game with so many complex things happening so quickly), but Dota matches can last for forty-five minutes or more. Although OpenAI Five can play with a terrifying level of aggression and precision, it also seems not to know how to use techniques that will pay off in the much longer term.[15] Like the simple Karate Kid bot that employs the Crane Kick too early, it tends to use up a character's most powerful attacks early on rather than saving them for later, when they will count the most.

This failure to plan ahead shows up fairly often. In level 2 of Super Mario Bros., there is an infamous ledge, the bane of all game-playing algorithms. This ledge has lots of shiny coins on it! By the time they get to level 2, AIs usually know coins are good. The AIs also usually know that they have to keep moving to the right so they can reach the end of the level before time runs out. But if the AI jumps onto the ledge, it then has to go backwards to get down off the ledge. The AIs have never had to go backwards before. They can't figure it out, and they get stuck on the ledge until time runs out. "I literally spent about six weekends and thousands of hours of CPU on the problem," said Tom Murphy, who eventually got past the ledge with some improvements to his AI's skills at long-term planning.[16]

Text generation is another place where the short memory of AI can be a problem. For example, Heliograf, the journalism algorithm that translates individual lines of a spreadsheet into sentences in a formulaic sports story, works because it can write each sentence more or less independently. It doesn't need to remember the entire article at once.

Language-translating neural networks, like the kind that power Google Translate, don't need to remember entire paragraphs, either. Sentences, or even parts of sentences, can usually be individually translated from one language to another without any memory of the previous sentence. When there is some kind of long-term dependence, such as an ambiguity that might have been resolved with information from a previous sentence, the AI usually can't make use of it.

Other kinds of tasks make AI's terrible memory even more obvious. One example is algorithmically generated stories. There's a reason AI doesn't write books or TV shows (though people are, of course, working on this).

If you're ever wondering whether a bit of text was written by a machine learning algorithm or a human (or at least heavily curated by a human), one way to tell is to look for major problems with memory. As of 2019, only some AIs are starting to be able to keep track of long-term information in a story — and even then, they'll tend to lose track of some bits of crucial information.

Many text-generating AIs can only keep track of a few words at a time. For example, here's what a **recurrent neural network (RNN)** wrote after it was trained on nineteen thousand descriptions of people's dreams from dreamresearch.net:

```
I get up and walk down the hall to his house and see a
bird in the very narrow drawer and it is a group of
people in the hand doors. At home like an older man is
going to buy some keys. He looks at his head with a
cardboard device and then my legs are parked on the
table.
```

Now, dreams are notoriously incoherent, switching settings and mood and even characters midstream. These neural-net dreams, however, don't maintain coherence for more than a sentence or so — sometimes considerably less. Characters who are never introduced are referred to as if they had been there all along. The whole dream forgets where it is. Individual phrases may make sense, and the rhythm of the words sounds okay if you don't pay attention to what's going on. Matching the surface qualities of human speech while lacking any deeper meaning is a hallmark of neural-net-generated text.

On the next page is another example, this time a recipe, where it's even easier to see the effects of memory limitation. This recipe was generated by the same recurrent neural network, or machine learning algorithm, that generated the recipes on pages 39–40. (As you can see, this is the one that learned from a variety of recipes, including, apparently, recipes for black pudding, a type of blood sausage.) This neural network builds a recipe letter by letter, looking at the letters it's already generated to decide which one comes next. But each extra letter that it looks at requires more memory, and there's only so much memory available on the computer that's running it. So to make the memory demands manageable, the neural network looks only at the most recent characters, a few at a time. For this particular algorithm and my computer, the largest memory I could give it was sixty-five characters. So every time it had to come up with the next letter of the recipe, it only had information about the previous sixty-five characters.* You can tell where in the recipe it ran out of memory and forgot it was making a chocolate dessert — about when it decided to add black pepper and whatever "rice cream" is.

This memory limitation is beginning to change. Researchers are working on making recurrent neural networks that can look at short-term *and*

---

* It also had a tiny bit of long-term memory where it could keep track of information for longer than that sixty five-character window, but that amount of memory was too tiny to store an entire ingredients list. In machine learning terms, that makes this algorithm a **Long Short-Term Memory (LSTM)** neural network rather than a plain RNN.

The struggle is going to be to remember what it's already written. It sees only 65 characters at a time: from beginning to the first semicolon. Can it at least stick with sweet vs savory?

The format is easy. Start with a title, then do a category, then ingredients, then directions. Predictable. It gets this every time. Predictable things are easy.

Chocolate Butterbroth Black Pudding,

Okay, I would have gone with "desserts" here. Let's try to stay on track.

cheese/eggs

This is made from blood. An interesting start.

Oh good, cocoa! We haven't forgotten about the chocolate yet. It's just barely in the 65-character memory.

4 oz cocoa; finely ground
1 teaspoon butter
1/2 cup milk
1/4 teaspoon pepper
1/4 cup rice cream, chopped
1 lb cream
1 sesame peel

Wait, it's already...well, good to make sure, I guess.

Why would we have 1 sesame anything? Clear error here. And peeling it seems tedious.

Sesame brings us dangerously close to savory territory.

----DATE HOLY----

Capital letters are a particular challenge because they're treated as completely unrelated to lower case. Neural net has to learn these independently from scratch with very few examples.

Frosting! Date Holy appears to be frosting. It must have lifted this from a cake recipe. We are back in dessert territory.

{ 1 large egg
1 powdered sugar serving barme
1/4 cup butter or margarine, melted

These weren't in the ingredients list, which has already passed out of memory by now. It got chocolate, but it's guessing.

Uh-oh. Ambiguity. "Until golden brown" could be sweet or savory. "Bubbly" tips the balance and now... Garlic. Game over.

Brown sugar, chocolate; baking powder, beer, lemon juice and salt in chunk in greased 9 x 2 inch cake.
Chill until golden brown and bubbly.
Place serve garlic half by pieoun on top to make more use bay.
Place in frying pan in preheated oven.
Sprinkle with fresh parsley for cooking.

Seems a bit small. And shouldn't this be a pan?

Oh no. It's completely lost track. Not enough info to figure out what's going on. Is this soup? Stir-fry? No good high-probability options. Even its spelling suffers as it flails.

Eating dish to hect in pot of the oil, pull over half-and half. Place in a bowl. Pat thetings on a strip of calaparo and wanned cooking in butter by cooking the seasoning. Sprinkle with onions. & Pull when bubbles and carrot is cooked, about 5 minutes. On a 15 inch hour a blender or wax paper mix to by dried roughes to boil (which is discovered.)

At least it remembered to close its parenthesis. There's probably a neuron just keeping track of parentheses.

Yield: 1 cake

Neural net at least knows how to wrap up a recipe. We were making cake, right? Let's say it's cake.

Sometimes it knows it should end the recipe soon because we're adding frosting or serving with something. With this chaos, it must have just guessed. Sometimes these recipes go on for pages, the neural net having no idea how long it has been.

long-term features when predicting the next letters in a text. The idea is similar to the algorithms that look at small-scale features in images first (edges and textures, for example), then zoom out to look at the big picture. These strategies are called **convolution**. A neural network that uses convolution (and that is also hundreds of times larger than the one I trained on my laptop) can keep track of information long enough to remain on topic. The following recipe is from a neural network called GPT-2, which OpenAI trained on a huge selection of webpages, and which I then fine-tuned by training it on all kinds of recipes.

Chunk Cake

cakes, deserts

8 cup flour

4 lb butter; room temperature

2 ¼ cup corn syrup; divided

2 eggs; pureed and cooled

1 teaspoon cream of tartar

½ cup m&m's

8 oz chunky whites

1 chocolate sifted

Cream 2 ¼ cups of flour at medium speed until thickened.

Lightly grease and flour two greased and waxed paper-lined box ingredients; combine flour, syrup, and eggs. Add cream of tartar. Pour into a gallon-size loaf-pan. Bake at 450 degrees for 35 minutes. Meanwhile, in large bowl, combine syrup, whites, and chocolate; stir which until thoroughly mixed. Cool pan. Pour 2 tb chocolate mixture over whole cake. Refrigerate until serving time.

Yield: 20 servings

With its memory improved by convolution, the GPT-2 neural net remembers to use most of its ingredients, and even remembers that it's supposed to be making cake. Its directions are still somewhat improbable — plain flour won't thicken no matter how long you cream it, and the flour/syrup/egg mixture is unlikely to turn into cake, even with the addition of cream of tartar. It's still an impressive improvement compared to the Chocolate Butterbroth Black Pudding.

Here's another example from GPT-2, this time its attempt at writing Harry Potter fan fiction. The algorithm was able to keep track of which characters were in the scene and even remember recurring motifs—in this case, remembering that there was already a snake on Snape's head.

> Snape: I understand.
> [A snake appears and Snape puts it on his head and it appears to do the talking. It says 'I forgive you.']
> HARRY: You can't go back if you don't forgive.
> Snape: [sighing] Hermione.
> HARRY: Okay, listen.
> Snape: I want to apologize to you for getting angry and upset over this.
> HARRY: It's not your fault.
> HARRY: That's not what I meant to imply.
> [Another snake appears then it says 'And I forgive you.']
> HERMIONE: And I forgive you.
> Snape: Yes.

Another strategy for dealing with memory limits is to group basic units together so the neural network can achieve coherence while remembering fewer things. Rather than remembering sixty-five letters, it might remember sixty-five entire words, or even sixty-five plot elements. If I had restricted my neural network to a specially crafted set of required ingredients and allowable ranges—as a team at Google did when trying to design a new gluten-free chocolate chip cookie—it would have produced valid recipes every time.[17] Unfortunately, Google's result, though more cookielike than anything my algorithm could have produced, was reportedly still terrible.[18]

IS THERE A SIMPLER WAY OF SOLVING THIS PROBLEM?

This leads us to one of the final things that determines whether a problem is a good one for AI (although it doesn't determine whether people will try to use AI to solve the problem anyway): is AI really the simplest way of solving it?

Some problems were tough to make progress on before we had big AI models and lots of data. AI revolutionized image recognition and language translation, making smart photo tagging and Google Translate ubiquitous. Those problems are hard for people to write down general rules for, but an AI approach can analyze lots of information and form its own rules. Or an AI can look at one hundred characteristics of phone customers who switched to a different provider, then figure out how to guess which customers are likely to switch in the future. Maybe the volatile customers are young, live in areas with poorer than average coverage, and have been customers for less than six months.

The danger, however, is misapplying a complex AI solution to a situation that would be better handled by a bit of common sense. Maybe the customers who leave are the ones on the weekly cockroach delivery plan — that plan is *terrible*.

LET THE AI DRIVE?

What about self-driving cars? There are many reasons why this is an attractive problem for AI. We would love to automate driving, of course — many people find it tedious or at times even impossible. A competent AI driver would have lightning-fast reflexes, would never weave or drift in its lane, and would never drive aggressively. In fact, self-driving cars tend to sometimes be *too* timid and have trouble merging with rush-hour traffic or turning left on a busy road.[19] The AI would never get tired, though, and could take the wheel for endless hours while the humans nap or party.

We can also accumulate lots of example data as long as we can afford to pay human drivers to drive around for millions of miles. We can easily build virtual driving simulations so that the AI can test and refine its strategies in sped-up time.

The memory requirements for driving are modest, too. This moment's steering and velocity don't depend on things that happened five minutes ago. Navigation takes care of planning for the future. Road hazards like pedestrians and wildlife come and go in a matter of seconds.

And finally, controlling a self-driving car is so difficult that we don't have other good solutions. AI is the solution that's gotten us the furthest so far.

Yet it is an open question whether driving is a narrow enough problem to be solved with today's AI or whether it will require something more like the human-level artificial general intelligence (AGI) I mentioned earlier. So far, AI-driven cars have proved themselves able to drive millions of miles on their own, and some companies report that a human needed to intervene on test drives only once every few thousand or so miles. It's that rare need for intervention, however, that's proving tough to eliminate fully.

Humans have needed to rescue the AIs of self-driving cars from a variety of situations. Usually companies don't disclose the reasons for these so-called disengagements, only the number of them, which is required by law in some places. This may be in part because the reasons for disengagement can be frighteningly mundane. In 2015 a research paper[20] listed some of them. The cars in question, among other things,

- saw overhanging branches as an obstacle,
- got confused about which lane another car was in,
- decided that the intersection had too many pedestrians for it to handle,
- didn't see a car exiting a parking garage, and
- didn't see a car that pulled out in front of it.

A fatal accident in March 2018 was the result of a situation like this — a self-driving car's AI had trouble identifying a pedestrian, classifying her first as an unknown object, then as a bicycle, and then finally, with only 1.3 seconds left for braking, as a pedestrian. (The problem was further confounded by the fact that the car's emergency braking systems were disabled in favor of alerting the car's backup driver, yet the system was not designed to actually alert the backup driver. The backup driver had also spent many, many hours riding with no intervention needed, a situation that would make the vast majority of humans less than alert.)[21] A fatal accident in 2016 also happened because of an obstacle-identification error — in this case, a self-driving car failed to recognize a flatbed truck as an obstacle (see the box on the next page).

In 2016 there was a fatal accident when a driver used Tesla's autopilot feature on city streets instead of the highway driving that it had been intended for. A truck crossed in front of the car, and the autopilot's AI failed to brake—it didn't register the truck as an obstacle that needed to be avoided. According to analysis by Mobileye (who designed the collision-avoidance system), because their system had been designed for highway driving, it had only been trained to avoid rear-end collisions. That is, it had only been trained to recognize trucks from behind, not from the side. Tesla reported that when the AI detected the truck, it recognized it as an overhead sign and decided it didn't need to brake.[22]

truck!!   ... road sign??

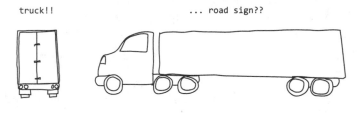

That's not to mention the more unusual situations that can occur. When Volkswagen tested its AI in Australia for the first time, they discovered it was confused by kangaroos. Apparently it had never before encountered anything that hopped.[23]

Given the sheer variety of things that can happen on a road—parades, escaped emus, downed electrical lines, lava, emergency signs with unusual instructions, molasses floods, and sinkholes—it's inevitable that something will occur that an AI never saw in training. It's a tough problem to make an AI that can deal with something completely unexpected—that would know that an escaped emu is likely to run wildly around while a sinkhole will stay put and to understand intuitively that just because lava flows and pools sort of like water does, it doesn't mean you can drive through a puddle of it.

Car companies are trying to adapt their strategies to the inevitability of mundane glitches or freak weirdness on the road. They're looking into limiting self-driving cars to closed, controlled routes (this doesn't necessarily solve the emu problem; they are wily) or having self-driving trucks caravan behind a lead human driver. In other words, the compromises are leading us toward solutions that look very much like mass public transportation.

Autonomy levels of self-driving cars

0. No Automation	Fixed-speed cruise control, at most. A Model T Ford qualifies. You're driving, end of story.
1. Driver Assistance	Adaptive cruise control or lane-keeping. Most modern cars have this. Some part of you is driving.
2. Partial Automation	2 or more things from level 1 work together. Car can maintain distance AND follow the road. Driver still must be ready to take over.
3. Conditional Automation	Car can drive by itself in some conditions. Cars with traffic jam mode; highway mode. Driver rarely needed but must be READY.
4. High Automation	Car doesn't need a driver on a controlled route. Sometimes, driver can go in back and nap. On other routes, still needs a driver.
5. Full Automation	Car never needs a driver. Car might not even have wheels and pedals. Go back to sleep. Car has it all under control.

As of right now, when the AIs get confused, they disengage — that is, they suddenly hand control back to the human behind the wheel. Automation level 3, conditional automation, is the highest level of car autonomy commercially available — in Tesla's autopilot mode, for example, the car can drive for hours unguided, but a human driver can be called to take over at any moment. The problem with this level of automation is that the human had better be behind the wheel and paying attention, not in the back seat decorating cookies. And humans are very, very bad at being alert after boring hours of idly watching the road. Human rescue is often a decent option for bridging the gap between the AI performance we have and the performance we need, but humans are pretty bad at rescuing self-driving cars.

So making self-driving cars is at once an attractive and very difficult AI problem. To get mainstream self-driving cars, we may need to make compromises (like creating controlled routes and sticking with automation level number 4), or we may need AI that's significantly more flexible than the AI we have now.

In the next chapter, we'll look at the types of AI that are behind things like self-driving cars — modeled after brains, evolution, and even the game of call my bluff.

CHAPTER 3

How does It actually learn?

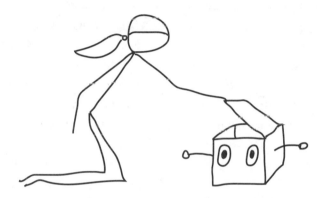

Remember that in this book I'm using the term *AI* to mean "machine learning programs." (Refer to the handy chart on page 8 for a list of stuff that I am or am not considering to be AI. Sorry, person in a robot suit.) A machine learning program, as I explained in chapter 1, uses trial and error to solve a problem. But how does that process work? How does a program go from producing a jumble of random letters to writing recognizable knock-knock jokes, all without a human telling it how words work or what a joke even is?

There are lots of different methods of machine learning, many of which have been around for decades, often long before people started calling them AI. Today, these technologies are combined or remixed or made ever more powerful by faster processing and bigger datasets. In this chapter we'll look at a few of the most common types, peeking under the hood to see how they learn.

NEURAL NETWORKS

These days, when people talk about AI, or **deep learning,** what they're often referring to are **artificial neural networks (ANNs)**. (ANNs have also been known as **cybernetics**, or **connectionism**.)

There are lots of ways to build artificial neural networks, each meant for a particular application. Some are specialized for image recognition, some for language processing, some for generating music, some for optimizing the productivity of a cockroach farm, some for writing confusing jokes. But they're all loosely modeled after the way the brain works. That's why they're called artificial neural networks — their cousins, **biological neural networks**, are the original, far more complex models. In fact, when programmers made the first artificial neural networks, in the 1950s, the goal was to test theories about how the brain works.

In other words, artificial neural networks are imitation brains.

They're built from a bunch of simple chunks of software, each able to perform very simple math. These chunks are usually called **cells** or **neurons**, an analogy with the neurons that make up our own brains. The power of the neural network lies in how these cells are connected.

Now, compared to actual human brains, artificial neural networks aren't that powerful. The ones I use for a lot of the text generation in this book have as many neurons as…a worm.

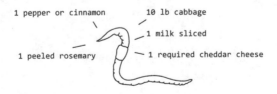

Unlike a human, the neural net is at least able to devote its entire one-worm-power brain to the task at hand (if I don't accidentally distract it with extraneous data). But how can you solve problems using a bunch of interconnected cells?

The most powerful neural networks, the ones that take months and tens of thousands of dollars' worth of computing time to train, have far more neurons than my laptop's neural net, some even exceeding the neuron count of a single honeybee. Looking at how the size of the world's largest neural networks has increased over time, a leading researcher estimated in 2016 that artificial neural networks might be able to approach the number of neurons in the human brain by around 2050.[1] Will this mean that AI will approach the intelligence of a human then? Probably not even close. Each neuron in the human brain is much more complex than the neurons in an artificial neural network—so complex that each human neuron is more like a complete many-layered neural network all by itself. So rather than being a neural network made of eighty-six billion neurons, the human brain is a neural network made of eighty-six billion neural networks. And there are far more complexities to our brains than there are to ANNs, including many we don't fully understand yet.

THE MAGIC SANDWICH HOLE

Let's say, hypothetically, that we have discovered a magic hole in the ground that produces a random sandwich every few seconds. (Okay, this is *very* hypothetical.) The problem is that the sandwiches are very, very random. Ingredients include jam, ice cubes, and old socks. If we want to find the good ones, we'll have to sit in front of the hole all day and sort them.

Alas, ear wax

But that's going to get tedious. Good sandwiches are only one in a thousand. However, they *are* very, very good sandwiches. Let's try to automate the job.

I can help!

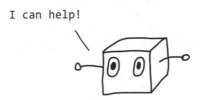

To save ourselves time and effort, we want to build a neural network that can look at each sandwich and decide whether it's good. For now, let's ignore the problem of how to get the neural network to recognize the ingredients the sandwiches are made of—that's a really hard problem. And let's ignore the problem of how the neural network is going to pick up each sandwich. That's also really, really hard—not just recognizing the motion of the sandwich as it flies from the hole but also instructing a robot arm to grab a slim paper-and-motor-oil sandwich or a thick bowling-ball-and-mustard sandwich. Let's assume, then, that the neural net knows what's in each sandwich and that we've solved the problem of physically moving the sandwiches. It just has to decide whether to save this sandwich for human consumption or throw it into the recycling chute. (We're also going to ignore the mechanism of the recycling chute — let's say it's another magic hole.)

This reduces our task to something simple and narrow — as we discovered in chapter 2, that makes it a good candidate for automation with a machine learning algorithm. We have a bunch of inputs (the names of the ingredients), and we want to build an algorithm that will use them to figure out our single output, a number that indicates whether the sandwich is good. We can draw a simple "black box" picture of our algorithm, and it looks like this:

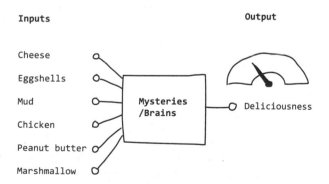

We want the "deliciousness" output to change depending on the combination of ingredients in the sandwich. So if a sandwich contains eggshells and mud, our black box should do this:

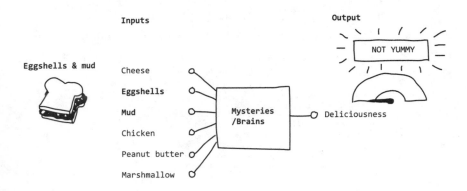

But if the sandwich contains chicken and cheese, it should do this instead:

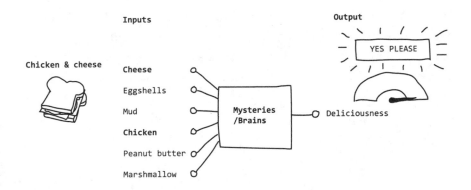

Let's look at how things are hooked up inside the black box.

First, let's make it simple. We hook up all the inputs (all the ingredients) to our single output. To get our deliciousness rating, we add each ingredient's contribution. Clearly each ingredient should not contribute equally — the presence of cheese would make the sandwich more delicious, while the presence of mud would make the sandwich less delicious. So each ingredient gets a different weight. The good ones get a weight of 1, while the ones we want to avoid get a weight of 0. Our neural network looks like this:

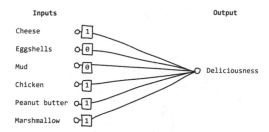

Let's test it with some sample sandwiches. Suppose the sandwich contains mud and eggshells. Mud and eggshells both contribute a 0, so the deliciousness rating is 0 + 0 = 0.

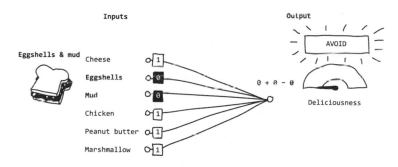

But a peanut-butter-and-marshmallow sandwich will get a rating of 1 + 1 = 2. (Congratulations! You have been blessed with that New England delicacy, the fluffernutter.)

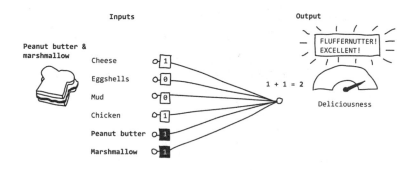

With this neural network configuration, we successfully avoid all the sandwiches that contain only eggshells, mud, and other inedible things. But this simple one-layer neural network is not sophisticated enough to recognize that some ingredients, while delicious on their own, are not delicious in combination with certain others. It's going to rate a chicken-and-marshmallow sandwich as delicious, the equal of the fluffernutter. It's also susceptible to something we'll call the **big sandwich bug**: a sandwich that contains mulch might still be rated as tasty if it contains enough good ingredients to cancel out the mulch.

To get a better neural network, we're going to need another layer of cells.

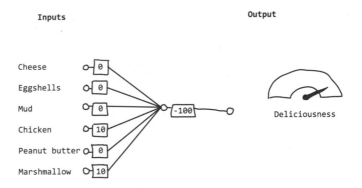

Here's our neural network now. Each ingredient is connected to our new layer of cells, and each cell is connected to the output. This new layer is called a **hidden layer**, because the user only sees the inputs and the outputs. Just as before, each connection has its own weight, so it affects our final deliciousness output in different ways. This isn't deep learning yet (that would require even more layers), but we're getting there.

DEEP LEARNING

Adding hidden layers to our neural network gets us a more sophisticated algorithm, one that's able to judge sandwiches as more than the sum of their ingredients. In this chapter, we've only added one hidden layer, but real-world neural networks often have several. Each new layer means a new way to combine the insights from the previous layer—at higher and higher levels of complexity, we hope. This approach—lots of hidden layers for lots of complexity—is known as **deep learning**.

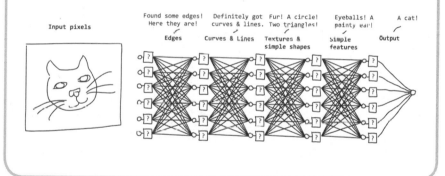

With this neural network, we can finally avoid bad ingredients by connecting them to a cell that we'll call the punisher. We'll give that cell a huge negative weight (let's say –100) and connect everything bad to it with a weight of 10. Let's make the first cell the punisher and connect the mud and eggshells to it. Here's what that looks like:

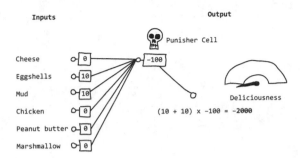

Now, no matter what happens in the other cells, a sandwich is likely to fail if it contains eggshells or mud. Using the punisher cell, we can beat the big sandwich bug.

We can do other things with the rest of the cells — like finally make a neural network that knows which ingredient combos work. Let's use the second cell to recognize chicken-and-cheese-type sandwiches. We'll refer to it as the deli sandwich cell. We connect chicken and cheese to it with weights of 1 (we'll also do this with ham and turkey and mayo) and connect everything else to it with weights of 0. And this cell gets connected to the output with a modest weight of 1. The deli sandwich cell is a good thing, but if we get too excited about it and assign it a very high weight, we'll be in danger of making the punisher cell less powerful. Let's look at what this cell does.

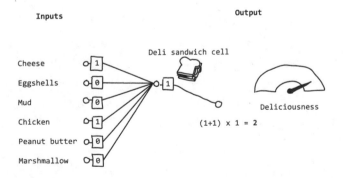

A chicken-and-cheese sandwich will cause this cell to contribute a cheerful 1 + 1 = 2 to the final output. But adding marshmallow to the chicken-and-cheese sandwich doesn't hurt it at all, even though it makes a pretty objectively less delicious sandwich. To fix that, we'll need other cells that specifically look for and punish incompatibilities.

Cell 3, for example, might look for the chicken-marshmallow combination (let's call it the cluckerfluffer) and severely punish any sandwich that contains it. It would be hooked up like this:

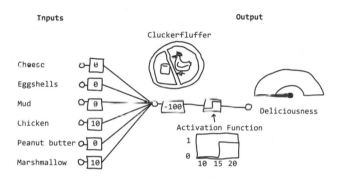

Cell 3 returns a devastating (10 + 10) × -100 = –2000 to any sandwich that dares to combine chicken and marshmallow. It's acting like a very specialized punisher cell, designed specifically to punish chicken and marshmallow. Notice that I've shown an extra part of the cluckerfluffer cell here, called the **activation function**, because without it, the cell will punish *any* sandwich that contains chicken *or* marshmallow. With a threshold of 15, the activation function stops the cell from turning on when just chicken (10 points) or marshmallow (10 points) is present — it will return a neutral 0. But if *both* are present (10 + 10 = 20 points), the threshold of 15 is exceeded, and the cell turns on. *Boom!* The activated cell punishes any combination of ingredients that exceeds its threshold.

Cluckerfluffer

With all the cells connected in similarly sophisticated configurations, we have a neural net that can sort out the best sandwiches the magic hole has to offer.

THE TRAINING PROCESS

So now we know what a well-configured sandwich-picking neural network might look like. But the point of using machine learning is that we don't have to set up the neural network by hand. Instead, it should be able to configure *itself* into something that does a great sandwich-picking job. How does this training process work?

Let's go back to a simple two-layer neural network. At the beginning of the training process, it's starting completely from scratch, with random weights for each ingredient. Chances are it's very, very bad at rating sandwiches.

We'll need to train it with some real-world data — some examples of the correct way to rate a sandwich, as demonstrated by real humans. As the neural net rates each sandwich, it needs to compare its ratings against those of a panel of cooperative sandwich judges. Note: never volunteer to test the early stages of a machine learning algorithm.

For this example, we'll go back to the very simple neural network. Remember, since we're trying to train it from scratch, we're ignoring all our prior knowledge about what the weights should be, and starting from random ones. Here they are:

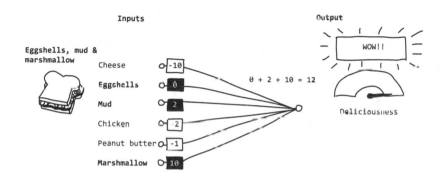

It *hates* cheese. It *loves* marshmallow. It's rather fond of mud. And it can take or leave eggshells.

The neural net looks at the first sandwich that pops out of the magic sandwich hole and using its (terrible) judgment, gives it a score. It's a marshmallow, eggshell, and mud sandwich, so it gets a score of 10 + 0 + 2 = 12. Wow! That's a really, really great score!

It presents the sandwich to the panel of human judges. Harsh reality: it's not a popular sandwich.

Now comes the part where the neural net has a chance to improve: it looks at what would have happened if its weights were slightly different. From this one sandwich, it doesn't know what the problem is. Was it too excited about the marshmallow? Are eggshells not neutral but maybe even a teensy bit bad? It can't tell. But if it looks at a batch of ten sandwiches, the scores it gave them, and the scores the human judges gave them, it can discover that if it had in general given mud a lower weight, lowering the score of any sandwich that contains mud, its scores would match those of the human judges a bit better.

With its newly adjusted weights, it's time for another iteration. The neural net rates another bunch of sandwiches, compares its scores against those of the human judges, and adjusts its weights again. After thousands more iterations and tens of thousands of sandwiches, the human judges are very, very sick of this, but the neural network is doing a lot better.

There are plenty of pitfalls in the way of progress, though. As I mentioned above, this simple neural network only knows if particular ingredients are generally good or generally bad, and it isn't able to come up with a nuanced idea of which combinations work. For that, it needs a more sophisticated structure, one with hidden layers of cells. It needs to evolve punishers and deli sandwich cells.

Another pitfall that we'll have to be careful of is the issue of **class imbalance**. Remember that only a handful of every thousand sandwiches from the sandwich hole are delicious. Rather than go through all the trouble of figuring out how to weight each ingredient, or how to use them in combination, the neural net may realize it can achieve 99.9 percent accuracy by rating each sandwich as terrible, no matter what.

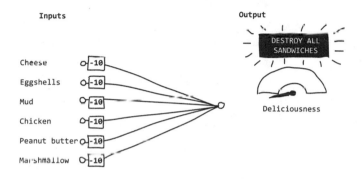

To combat class imbalance, we'll need to prefilter our training sandwiches so that there are approximately equal proportions of sandwiches that are delicious and awful. Even then, the neural net might not learn about ingredients that are usually to be avoided but delicious in very specific circumstances. Marshmallow might be an example of such an ingredient — awful with most of the usual sandwich ingredients but delicious in a fluffernutter (and maybe with chocolate and bananas). If the neural net doesn't see fluffernutters in training, or sees them very rarely, it may decide that it can achieve pretty good accuracy by rejecting anything that contains marshmallow.

Class imbalance–related problems show up all the time in practical applications, usually when we ask AI to detect a rare event. When people try to predict when customers will leave a company, they have a lot more examples of customers who stay than customers who leave, so there's a danger the AI will take the shortcut of deciding that all customers will stay forever. Detecting fraudulent logins and hacking attacks has a similar problem, since actual attacks are rare. People also report class imbalance problems in medical imaging, where they may be looking for just one abnormal cell among hundreds—the temptation is for the AI to shortcut its way to high accuracy just by predicting that all cells are healthy. Astronomers also run into class imbalance problems when they use AI, since many interesting celestial events are rare—there was a solar-flare-detecting program that discovered it could achieve near 100 percent accuracy by predicting zero solar flares, since these were very rare in the training data.[2]

WHEN CELLS WORK TOGETHER

In the sandwich-sorting example above, we saw how a layer of cells can increase the complexity of the tasks a neural network can perform. We built a deli sandwich cell that responded to combinations of deli meats and cheeses, and we built a cluckerfluffer cell that punished any sandwich that tried to use chicken and marshmallow in combination. But in a neural network that trains itself, using trial and error to adjust the connections between cells, it's usually a lot harder to identify each cell's job. Tasks tend to be spread among several cells — and in the case of some cells, it's difficult or impossible to tell what tasks they accomplish.

To explore this phenomenon, let's look at some of the cells of a fully trained neural net. Built and trained by researchers at OpenAI,[3] this particular neural net looked at more than eighty-two million Amazon product reviews letter by letter and tried to predict which letter would come next. This is another recurrent neural network, the same general sort as the one that generated the knock-knock jokes, ice cream flavors, and recipes listed in chapters 1 and 2. This one's larger — it has approximately as many neurons as a jelly fish. Here are a few examples of reviews it generated:

> This is a great book that I would recommend to anyone who loves the great story of the characters and the series of books.

> I love this song. I listen to it over and over again and never get tired of it. It is so addicting. I love it!!

> This is the best product I have ever used to clean my shower stall. It is not greasy and does not strip the water of the water and stain the white carpet. I have been using it for a few years and it works well for me.

> These workout DVDs are very useful. You can cover your whole butt with them.

> I bought this thinking it would be good for the garage. Who has a lot of lake water? I was totally wrong. It was simple and fast. The night grizzly has not harmed it and we have had this for over 3 months. The guests are inspired and they really enjoy it. My dad loves it!

This particular neural net has an input for each letter or punctuation mark it could encounter (similar to the sandwich sorter, which had one input for each sandwich ingredient) and can look back at the past few letters and punctuation marks. (It is as if the sandwich rater's scoring depended a bit on the last few sandwiches it had seen — maybe it can keep track of whether we might be sick of cheese sandwiches and adjust the next cheese sandwich's rating accordingly.) And rather than having a single output, as the sandwich sorter does, the review-writing neural net has a lot of them, one output for each letter or punctuation mark that it could choose as most likely to come next in the review. If it sees the sequence "I own twenty eggbeaters and this is my very favorit," then the letter *e* will be the most likely next choice.

Based on the outputs, we can take a look at each cell and see when it's "active," letting us make an educated guess about what its function is. In our sandwich-sorter example above, the deli sandwich cell would be active when it sees lots of meat and cheese and inactive when it sees socks or marbles or peanut butter. However, most of the neurons in the Amazon product-review neural net are going to be nowhere near as interpretable as deli cells and punisher neurons. Instead, most of the rules the neural net comes up with are going to be unintelligible to us. Sometimes we can guess what a cell's function will be, but far more frequently, we have no idea what it's doing.

Here's the activity of one of the product-review algorithm's cells (the 2,387th) as it generates a review (white = active, dark = inactive):

```
For me, this is one of the few albums of theirs I own
that actually made me an instant classic pop fan. I also
```

had a major problem with the audio with 10 new songs;
the execution of the vocals and editing was awful. The
next day, I was in a recording studio and I can't tell
you how many times I had to hit the play button to see
where the song was going.

This cell is contributing to the neural net's prediction of which letters come next, but its function is mysterious. It's reacting to certain letters, or certain combinations of letters, but not in a way that makes sense to us. Why was it really excited about the letters *um* in *album* but not the letters *al*? What is it actually doing? It's just one small piece of the puzzle working with a lot of other cells. Almost all the cells in a neural net are as mysterious as this one.

However, every once in a while, there will be a cell whose job is recognizable — a cell that activates whenever we're between a pair of parentheses or that activates increasingly strongly the longer a sentence gets.[4] The people who trained the product-review neural net noticed that it had one cell that was doing something they could recognize: it was responding to whether the review was positive or negative. As part of its task of predicting the next letter in a review, the neural net seems to have decided it was important to determine whether to praise the product or trash it. Here's the activation of the "sentiment neuron" on that same review. Note that a light color indicates high activation, which means it thinks the review is positive:

For me, this is one of the few albums of theirs I own
that actually made me an instant classic pop fan. I also
had a major problem with the audio with 10 new songs;
the execution of the vocals and editing was awful. The
next day, I was in a recording studio and I can't tell
you how many times I had to hit the "play" button to see
where the song was going.

The review starts out very positive, and the sentiment neuron is highly activated. Midway through, however, it switches tone, and the cell's activation level goes way down.

Here's another example of the sentiment neuron at work. It has low activity when the review is neutral or critical but quickly swings into high gear whenever it detects a change in sentiment:

```
The Harry Potter File, from which the previous one was
based (which means it has a standard size liner) weighs
a ton and this one is huge! I will definitely put it on
every toaster I have in the kitchen since, it is that
good. This is one of the best comedy movies ever made.
It is definitely my favorite movie of all time. I would
recommend this to ANYONE!
```

But it's less good at detecting sentiment in other kinds of text. Most people would not classify this passage from Edgar Allan Poe's "The Fall of the House of Usher" as positive in sentiment, but this particular neural net thinks it's mostly positive:

```
Overpowered by an intense sentiment of horror,
unaccountable yet unendurable, I threw on my clothes
with haste (for I felt that I should sleep no more
during the night,) and endeavoured to arouse myself
from the pitiable condition into which I had fallen, by
pacing rapidly to and fro through the apartment.
```

I guess a movie could overpower you by an intense sentiment of horror and be a good movie if that's what it was supposed to do.

Again, it's unusual to find a cell in a text-generating or text-analyzing algorithm that behaves as transparently as the sentiment neuron. The same goes for other types of neural networks — and that's too bad, since we'd love to be able to tell when they're making unfortunate mistakes and to learn from their strategies.

In image-recognizing algorithms, though, it's a bit easier to find cells whose jobs you can identify. There the inputs are the individual pixels of a particular image, and the outputs are the various possible ways to classify the image (dog, cat, giraffe, cockroach, and so on). Most image recognition algorithms have lots and lots of layers of cells in between — the hidden layers. And in most image recognition algorithms, there are cells or groups of cells whose functions we can identify if we analyze the neural net in the right way. We can look at the collections of cells that activate when they see particular things, or we can tweak the input image and see which changes make the cells activate most strongly.

DEEP DREAMING

Tweaking an image to make the neurons more excited about it is the technique used to make the famous Google DeepDream images where an image-identifying neural network turned ordinary images into landscapes full of trippy dog faces and fantastic conglomerations of arches and windows.

To make a DeepDream image, you start with a neural network that has been trained to recognize something—dogs, for example. Then you choose one of its cells and gradually change the image to make that cell increasingly more excited about it. If the cell is trained to recognize dog faces, then it will get more excited the more it sees areas in the image that look like dog faces. By the time you've changed the image to the cell's liking, it will be highly distorted and covered in dogs.

The smallest groups of cells seem to look for edges, colors, and very simple textures. They might report vertical lines, curves, or green grassy textures. In subsequent layers, larger groups of cells look for collections of edges, colors, and textures or for simple features. Some researchers at Google, for example, analyzed their GoogLeNET image recognition algorithm and found that it had several collections of cells that were looking specifically for floppy versus pointy ears on animals, which helped it distinguish dogs from cats.[5] Other cells got excited about fur or eyeballs.

Image-generating neural networks also have some cells that do identifiable jobs. We can do "brain surgery" on image-generating neural networks, removing certain cells to see how the generated image changes.[6] A group at MIT found that it could deactivate cells to remove elements from generated images. Interestingly, elements that the neural net deemed "essential" were more difficult to remove than others—for example, it was easier to remove curtains from an image of a conference room than to remove the tables and chairs.

Now let's look at another kind of algorithm, one you've probably interacted with directly if you've used the predictive-text feature of a smartphone.

MARKOV CHAINS

A **Markov chain** is an algorithm that can tackle many of the same problems as the recurrent neural network (RNN) that generated the recipes, ice cream flavors, Amazon reviews, and metal bands in this book. Like the RNN, it looks at what happened in the past (words previously used in a sentence or last week's weather, for example) and predicts what's most likely to happen next.

Markov chains are more lightweight than most neural networks and quicker to train. That's why the predictive-text function of smartphones is usually a Markov chain rather than an RNN.

However, a Markov chain gets exponentially more unwieldy as its memory increases. Most predictive-text Markov chains, for example, have memories that are only three to five words long. RNNs, by contrast, can have memories that are hundreds of words long — or even longer with the use of **LSTM** (long short-term memory) and convolution tricks. In chapter 2 we saw how important memory length is when short memory made an RNN lose track of important information. The same is true for Markov chains.

I trained a Markov chain with a dataset of Disney songs using a trainable predictive-text keyboard.[7] Training took only a few seconds as opposed to a few minutes for an RNN. But this Markov chain has a three-word memory. That is, the words it suggests are the ones it thinks are the most likely based on the previous three words in the song. When I had it generate a song, choosing only its top suggestion at every step, here is what it produced:

```
The sea)
under the sea)
under the sea)
under the sea)
```

```
under the sea)
under the sea)
under the sea)
```

It doesn't know how many times to sing "under the sea" because it doesn't know how many times it has already sung it.

If I start it out with the beginning of the song "Beauty and the Beast" ("Tale as Old as Time"), it quickly gets stuck again.

```
Tale as old as time
song as old as time
song as old as time
song as old as time
```

In several verses of "Beauty and the Beast," the words "tale as old as time" are immediately followed by the words "song as old as rhyme." But when this Markov chain is looking at the phrase "as old as," it doesn't know which of those two verses it's in the middle of writing.

I can get it out of its trap by choosing the *second* most probable word at every step. Then it writes this:

```
A whole world
bright young master
you're with all
ya think you're by wonder
by the powers
and i got downhearted
alone hellfire dark side
```

And choosing the third most probable word each time:

```
You think i can open up
where we'll see how you feel
it all my dreams will be mine
is something there before
she will be better time
```

These are a lot more interesting, but they don't make much sense. And songs — and poetry — are pretty forgiving when it comes to grammar, structure, and coherence. If I give the Markov chain a different dataset to learn, then its shortcomings become even more obvious.

Here's a Markov chain trained on a list of April Fool's Day pranks as it chooses the most probable next word at each step. (It never suggested punctuation, so the line breaks are my additions.)

```
The door knob off a door and put it back on backwards
    softly
Do nothing all day to a co of someone's ad in the paper
    for a garage sale at someone of an impending prank
Then do nothing all day to a co of someone's ad in the
    paper for a garage sale at . . .
```

A predictive-text Markov chain isn't likely to hold a conversation with a customer or write a story that can be used as a new video-game quest (both of which are things that people are trying to get RNNs to do one day). But one thing it can do is suggest likely words that might come next in a particular training set.

The people at Botnik, for example, use Markov chains trained on various datasets (Harry Potter books, *Star Trek* episodes, Yelp reviews, and more) to suggest words to human writers. The unexpected Markov chain suggestions often help the writers take their texts in weirdly surreal directions.

Rather than allowing the Markov chain and its short memory to try to choose the next word, I can let it come up with a bunch of options and present them to me — just as predictive text does when I'm composing a text message to someone.

Here's an example of what it looks like to interact with one of Botnik's trained Markov chains, this one trained on Harry Potter books:

```
≡                                    Predictive Writer


   Harry stared incredulously at dumbledore as he sat in
   a pool of |




   Source: Hp Attribution          Shuffle ⤫   Publish ⇧
```

the	his	her
them	a	him
it	what	harry's
parchment	sight	course
harry	magic	magical
green	panic	their

And here are some new April Fool's Day pranks I wrote with the help of the predictive text of a trained Markov chain:

```
Put plastic wrap pellets on your lips.
Arrange the kitchen sink into a chicken head.
Put a glow stick in your hand and pretend to sneeze on
    the roof.
Make a toilet seat into pants and then ask your car to
    pee.
```

For the sake of comparison, I also used a more complex, data-intensive RNN to generate April Fool's Day pranks. In this case, the RNN generated the entire prank, punctuation and all. However, there was still an element of human creativity involved—I had to sort through all the RNN-generated pranks looking for the funniest ones.

```
Make a food in the office computer of someone.
Hide all of the entrance to your office building
    if it only has one entrance.
Putting googly eyes on someone's computer mouse
    so that it won't work.
Set out a bowl filled with a mix of M&M's,
    Skittles, and Reese's Pieces.
Place a pair of pants and shoes in your ice
    dispenser.
```

You can conduct similar experiments with the predictive text included in most phone messaging apps. If you start with "I was born…" or "Once upon a time…" and keep clicking the phone's suggested words, you'll get a strange piece of writing straight from the innards of a machine learning algorithm. And because training a new Markov chain is relatively quick and easy, the text you get is specific to you. Your phone's predictive text and autocorrect Markov chains update themselves as you type, training themselves on what you write. That's why if you make a typo, it may haunt you for quite some time.

Google Docs may have fallen victim to a similar effect when users reported its autocorrect would change "a lot" to "alot" and suggested "gonna" instead of "going." Google was using a context-aware autocorrect

Did you mean:
spgheiit sauce?

that scanned the internet to decide which suggestions to make.[8] On the plus side, a context-aware autocorrect is able to spot typos that form real words (like "gong" typed intead of "going"), and add new words as soon as they become common. However, as any user of the internet knows, common usage rarely dovetails with the grammatically "correct" formal usage you'd want in a word processor's autocorrect feature. Although Google hasn't talked specifically about these autocorrect bugs, the bugs do tend to disappear after users report them.

RANDOM FORESTS

A **random forest algorithm** is a type of machine learning algorithm frequently used for prediction and classification — predicting customer behavior, for example, or making book recommendations or judging the quality of a wine — based on a bunch of input data.

To understand the forest, let's start with the trees. A random forest algorithm is made of individual units called decision trees. A **decision tree** is basically a flowchart that leads to an outcome based on the information we have. And, pleasingly, decision trees do kind of look like upside-down trees.

On the next page is a sample decision tree for, hypothetically, whether to evacuate a giant cockroach farm.

The decision tree keeps track of how we use information (ominous noises, the presence of cockroaches) to make decisions about how to handle the situation. Just as our sandwich decisions become more sophisti-

cated as the number of cells in our neural network increases, we can handle the cockroach situation with more nuance if we have a larger decision tree.

If the cockroach farm is strangely quiet, yet the roaches have not escaped, then there may be other explanations (perhaps even more unsettling) besides "they're all dead." With a larger tree we could ask whether there are dead cockroaches around, how smart the cockroaches are known to be, and whether the cockroach-crushing machines have been mysteriously sabotaged.

With lots and lots of inputs and choices, the decision tree can become hugely complex (or, to use the programming parlance of deep learning, very deep). It could become so deep that it encompasses every possible input, decision, and outcome in the training set, but then the chart would only work for the specific situations from the training set. That is, it would overfit the training data. A human expert could cleverly construct a huge decision tree that avoids overfitting and can handle most decisions without fixating on specific, probably irrelevant data. For example, if it was cloudy and cool the last time the cockroaches got out, a human is smart enough to know that having the same weather doesn't necessarily have anything to do with whether the cockroaches will escape again.

But an alternative approach to having a human carefully build a huge decision tree is to use the random forest method of machine learning. In much the same way as a neural network uses trial and error to configure the connections between its cells, a random forest algorithm uses trial and error to configure itself. A random forest is made of a bunch of tiny (that is, shallow) trees that each consider a tiny bit of information to make a couple of small decisions. During the training process, each shallow tree learns which information to pay attention to and what the outcome should be. Each tiny tree's decision probably won't be very good, because it's based on very limited information. But if all the tiny trees in the forest pool their decisions and vote on the final outcome, they will be much more accurate than any individual tree. (The same phenomenon holds true for human voters: if people try to guess how many marbles are in a jar, individually their guesses may be way off, but on average their guesses will likely be very close to the real answer.) The trees in a random forest can pool their decisions on all sorts of topics, coming up with an accurate picture of staggeringly complex scenarios. One recent application, for example, was sorting through hundreds of thousands of genomic patterns to determine which species of livestock was responsible for a dangerous E. coli outbreak.[9]

If we used a random forest to handle the cockroach situation, here's what a few of its trees might look like:

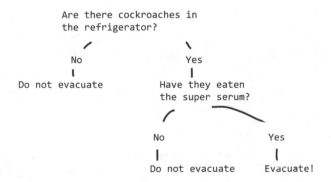

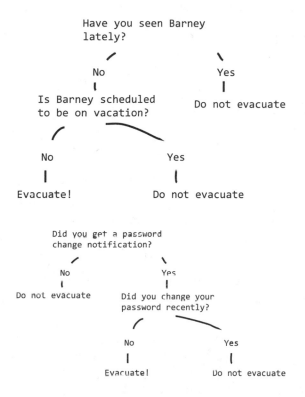

Now, each individual tree is only seeing a very small bit of the situation. There may be a perfectly reasonable explanation for why Barney isn't around — perhaps Barney has merely called in sick. And if the cockroaches have not actually eaten the super serum, that doesn't necessarily mean we're safe. Maybe the cockroaches have taken samples of the super serum and are even now brewing up a huge batch, enough for the 1.7 billion cockroaches in the facility.

But the trees are combining their individual hunches, and with Barney mysteriously missing, the serum gone, and your password mysteriously changed, the decision to evacuate may be a prudent one.

EVOLUTIONARY ALGORITHMS

AI refines its understanding by making a guess about a good solution, then testing it. All three machine learning algorithms above use trial and error to refine their own structures, producing the configuration of neurons, chains, and trees that lets them best solve the problem. The simplest methods of trial and error are those in which you always travel in the direction of improvement — often called **hill climbing** if you're trying to maximize a number (say, the number of points collected during a game of Super Mario Bros.) or **gradient descent** if you're trying to minimize a number (like the number of escaped cockroaches). But this simple process of getting closer to your goal doesn't always yield the best results. To visualize the pitfalls of simple hill climbing, imagine you're somewhere on a mountain (in deep fog) and trying to find its highest point.

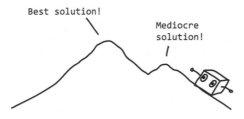

If you use a simple hill-climbing algorithm, you'll head uphill no matter what. But depending on where you start, you might end up stopping at the

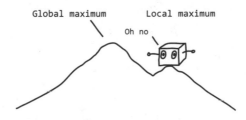

lowest peak — a **local maximum** — rather than the highest peak, the **global maximum**.

So there are more complex methods of trial and error designed to force you to try out more parts of the mountain, maybe doing a few test hikes in a few different directions before deciding where the most promising areas are. With those strategies, you might end up exploring the mountain more efficiently.

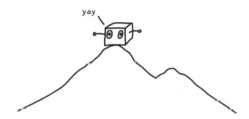

In machine learning terms, the mountain is called your **search space** — somewhere in that space is your goal (that is, somewhere on the mountain is the peak), and you're trying to find it. Some search spaces are **convex**, meaning that a basic hill-climbing algorithm will find you the peak each time. Other search spaces are much more annoying. The worst are the so-called **needle-in-the-haystack problems**, in which you might have very little clue how close you are to the best solution until the moment you stumble upon it. Searching for prime numbers is an example of a needle-in-the-haystack problem.

The search space of a machine learning algorithm could be anything. For example, the search space could be the shapes of parts that make up a

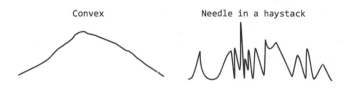

Convex

Needle in a haystack

walking robot. Or it could be the set of possible weights of a neural network, and the "peak" is the weights that help you identify fingerprints or faces. Or the search space could be the set of possible configurations of a random forest algorithm, and your goal is to find a configuration that's good at predicting a customer's favorite books — or whether the cockroach factory should be evacuated.

As we learned above, a basic search algorithm like hill climbing or gradient descent might not get you very far if the search space of possible neural net configurations is not very convex. So machine learning researchers sometimes turn to other, more complex trial-and-error methods.

One of these strategies takes its inspiration from the process of evolution. It makes a lot of sense to imitate evolution — after all, what is evolution if not a generational process of "guess and check"? If a creature is different from its neighbors in some way that makes it more likely to survive and therefore reproduce, then it will be able to pass its useful traits on to the next generation. A fish that can swim a tiny bit faster than other individuals of its species may be more likely to escape predators, and after a few generations of this, its fast-swimming offspring may be a bit more common than the descendants of slower-swimming fish. And evolution is a powerful, powerful process — one that has solved countless locomotion and information-processing problems, figured out how to extract food from sunlight and from hydrothermal vents, and figured out how to glow, fly, and hide from predators by looking like bird dung.

In **evolutionary algorithms**, each potential solution is like an organism. In each generation, the most successful solutions survive to reproduce, mutating or mating with other solutions to produce different — and, one hopes, better — children.

If you've ever struggled to solve a complex problem, it might be mind-boggling to think of each potential solution as a living being — eating, mating, whatever. But let's think about it in concrete terms. Let's say we're trying to solve a crowd-control problem: we have a hallway that splits into a fork, and we want to design a robot that can direct people to take one hall-way or the other.

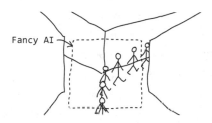

Fancy AI

The first thing we do is come up with the bits that the evolutionary algorithm can vary, deciding what about our robot we want to be constant and what the algorithm is free to play with. We could make these variable elements very limited, with a fixed body design, and just allow the program to change the way the robot moves around. Or we could allow the algorithm to build a body design completely from scratch, starting from random blobs. Let's say that the owners of this building are insisting on a humanlike robot design for sci-fi-aesthetic reasons. No messy jumble of crawling blocks (which is what an evolutionary algorithm's creatures tend to look like, given absolute freedom). Within a basic humanlike form, there's still a lot we could vary, but let's keep it simple and say that the algorithm will be allowed to vary the size and shape of a few basic body parts, with each one having a simple range of motion. In evolutionary terms, this is the robot's **genome**.

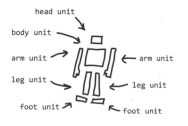

head unit
body unit
arm unit
leg unit
foot unit
arm unit
leg unit
foot unit

Robot Genome

Body part dimensions:
  Head unit: length, width, height
  Body unit: length, width, height
  ...
Behaviors:
  Default behavior
  When human present
  When human moves left
  When human moves right
  ...

The next thing we need to do is define the problem we're trying to solve in such a way that there's a single number we can optimize. In evolutionary terms, this number is the **fitness function**: a single number that will describe how fit an individual robot is for our task. Since we're trying to build a robot that can direct humans down one hallway or the other, let's say that we're trying to minimize the number of humans that take the left-hand fork. The closer that number is to zero, the higher the fitness.

We'll also need a simulation, because there's no way we're building thousands of robots to order or hiring people to walk down a hall thousands of times. (Not using real humans is also a safety consideration — for reasons that will be clear later.) So let's say it's a simulated hall in a world with simulated gravity and friction and other simulated physics. And of course we need simulated people with simulated behaviors, including walking, lines of sight, crowding, and various phobias, motivations, and levels of cooperativeness. The simulation itself is a really hard problem, so let's just say we've solved it already. (Note: in actual machine learning, it's never this easy.)

One handy way of getting a ready-made simulation that can train an AI is to use video games. That's partly why there are so many researchers training AIs to play Super Mario Bros. or old Atari games — these old video games are small, quick-to-run programs that can test various problem-solving skills. Just like human video-game players, though, AIs tend to find and exploit bugs in the games. More about this in chapter 5.

We let the algorithm randomly create our first generation of robots. They're...very random. A typical generation produces hundreds of robots, each with a different body design.

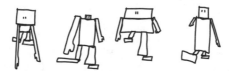

Then we test each robot individually in our simulated hallway. They don't do well. People walk right past them as they flop on the ground and flail ineffectually. Maybe one of them falls a bit more to the left than the others and blocks that hallway slightly, and a few of the more timid humans decide to take the right hallway instead. It scores slightly better than the other robots.

Now it's time to build the next generation of robots. First, we'll choose which robots are going to survive to reproduce. We could save just the very best robot, but that would make the population pretty uniform, and we wouldn't get to try out some other robot designs that might end up being better if evolution gets a chance to tweak them. So we'll save some of the best robots and throw out the rest.

Next, we have lots of choices about how the surviving robots are going to reproduce. They can't simply make identical copies of themselves, because we want them to be evolving toward something better. One option we have is **mutation**: pick a random robot and randomly vary something about it.

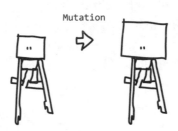

Another option we might decide to use is **crossover**: two robots produce offspring that are random combinations of the two parents.

We also have to decide how many offspring each robot can have (should the most successful robots have the most offspring?), which robots can cross with which other robots (or if we use crossover at all), and whether we're going to replace all the dead robots with offspring or with a few randomly generated robots. Tweaking all these options is a big part of building an evolutionary algorithm, and sometimes it's hard to guess which options — which **hyperparameters** — are going to work best.

Once we've built the new generation of robots, the cycle begins again as we test their crowd-controlling abilities in the simulation. More of them are now flopping over to the left because they're descended from that first marginally successful robot.

After many more generations of robots, some distinct crowd-control strategies start to emerge. Once the robots learn to stand up, the original "fall to the left and be kinda in the way" strategy has evolved into a "stand in the left hallway and be even more annoying" strategy. Another strategy also emerges — the "point vigorously to the right" strategy. But none of the strategies is perfectly solving our problem yet: each robot is still letting plenty of people leak into the left hallway.

After many more generations, a robot emerges that is very good at preventing people from entering the left hallway. Unfortunately, by a stroke of bad luck, it just so happens that the solution it found was "murder everyone." Technically that solution works because all we told it to do was minimize the number of people entering the left hallway.

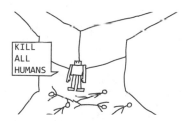

Because of a problem with our fitness function, evolution directed the algorithm toward a solution that we hadn't anticipated. Unfortunate shortcuts happen in machine learning all the time, although not usually this dramatically. (Fortunately for us, in real life, "kill all humans" is usually very impractical. Don't give autonomous algorithms deadly weapons is the message here.) Still, *this* is why we used simulated humans rather than real humans in our thought experiment.

We'll have to start over again, this time with a fitness function that, rather than minimizing the number of humans in the left-hand hallway, maximizes the number of humans who take the right-hand hallway.

Actually, we can take a (somewhat gory) shortcut and just change the fitness function rather than completely starting over. After all, our robots have learned many useful skills besides murdering people. They've learned to stand, detect people, and move their arms in a scary manner. Once our fitness function changes to maximizing the number of survivors who enter the right-hand hallway, the robots *should* quickly learn to forsake their murdering ways. (Recall that this strategy of reusing a solution from a different but related problem is called transfer learning.)

So we start with the group of murdering robots and sneakily swap the fitness function on them. Suddenly, murdering isn't working very well at all, and they don't know why. In fact, the robot that was the worst at murdering is now at the top of the heap, because some of its screaming victims managed to escape down the right-hand hallway. Over the next few generations, the robots quickly become ever worse at murdering.

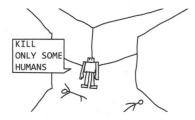

Eventually, maybe they only look like they might *want* to murder you, which would scare most humans into entering the right-hand hallway. By starting with a population of murderbots, we do restrict the path that evolution is likely to take. Had we started over instead, we might have evolved robots that stood at the end of the right-hand hallway and beckoned people or even robots whose hands evolved into signs that said FREE COOKIES. (The "free cookies" robot would be hard to evolve, though, because getting the sign merely partially right wouldn't work at all, and it would be hard to reward a solution that was only getting close. In other words, it's a needle-in-the-haystack solution.)

All murderbots aside, the most likely path that evolution would have taken is the "fall down and be in the way" robot getting ever more annoy-

ingly in the way. (Falling down is pretty easy to do, so if an evolved robot can solve a problem by falling down, it will tend to do that.) Through that path we may arrive at a robot that solves the problem perfectly by causing 100 percent of humans to enter the right-hand hallway (murdering none of them in the process). The robot looks like this:

Yes, we have evolved: a door.

That's the other thing about AI. It can sometimes be a needlessly complicated substitute for a commonsense understanding of the problem.

Evolutionary algorithms are used to evolve all kinds of designs, not just robots. Car bumpers that dissipate force when they crumple, proteins that bind to other medically useful proteins, flywheels that spin just so — these are all problems that people have used evolutionary algorithms to solve. The algorithm doesn't have to stick to a genome that describes a physical object, either. We could have a car or bicycle with a fixed design and a control program that evolves. I mentioned earlier that the genome can even be the weights of a neural network or the arrangement of a decision tree.

Different kinds of machine learning algorithms are often combined like this, each playing to its strength.

When we consider the huge array of life that has arisen on our planet via evolution, we get an idea of the magnitude of possibility that's available to us by using virtual evolution at a massively accelerated speed. Just as real-life evolution has managed to produce wonderfully complex creatures and allow them to take advantage of the weirdest, most specific food sources, evolutionary algorithms continue to surprise and delight us with their ingenuity. Of course, sometimes evolutionary algorithms can be a little *too* creative — as we'll see in chapter 5.

GENERATIVE ADVERSARIAL NETWORKS (GANS)

AIs can do amazing things with images, turning a summer scene into a winter one, generating faces of imaginary people, or changing a photo of someone's cat into a cubist painting. These showy image-generating, image-remixing, and image-filtering tools are usually the work of **GANs (generative adversarial networks)**. They're a subvariety of neural networks, but they deserve their own mention. Unlike the other kinds of machine learning in this chapter, GANs haven't been around very long — they were only introduced by Ian Goodfellow and other Université de Montréal researchers in 2014.[10]

The key thing about GANs is they're really two algorithms in one — two adversaries that learn by testing each other. One, the **generator**, tries to imitate the input dataset. The other, the **discriminator**, tries to tell the difference between the generator's imitation and the real thing.

To see why this is a helpful way of training an image generator, let's go through a hypothetical example. Suppose we want to train a GAN to generate images of horses.

The first thing we'll need is lots of example pictures of horses. If they all show the same horse in the same pose (maybe we're obsessed with that particular horse), the GAN will learn more quickly than if we give it a huge variety of colors and angles and lighting conditions. We can also simplify things by using a plain, consistent background. Otherwise the GAN will spend a long time trying to learn when and how to draw fences, grass, and parades. Most of the GANs that can generate photorealistic faces, flowers, and foods were given very limited, consistent datasets — pictures of just cat faces, for example, or bowls of ramen photographed only from the top. A GAN trained just on photos of tulip heads may produce very convincing tulips but will have no idea about other kinds of flowers or even any concept that tulips have leaves or bulbs. A GAN that can generate photorealistic human head shots won't know what's below the neck, what's on the back of the head, or even that human eyes can close. So this is all to say that if we're going to make a horse-generating GAN, we'll have better success if we make its world a very simple one and only give it pictures of horses photographed from the side against a plain white background. (Conveniently, this is also about the extent of my drawing ability.)

Now that we have our dataset (or, in our case, now that we've imagined one), we're ready to start training the two parts of the GAN, the generator and the discriminator. We want the generator to look at our set of horse pictures and figure out some rules that will let it make pictures similar to them. Technically what we are asking the generator to do is warp random noise into pictures of horses — that way, we can get it to generate not just one single horse picture but also a different horse for every random noise pattern.

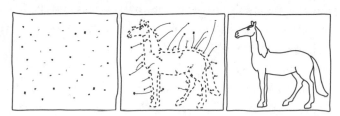

At the beginning of the training, though, the generator hasn't learned any rules about drawing horses. It starts with our random noise and does something random to it. As far as it knows, that is how you draw a horse.

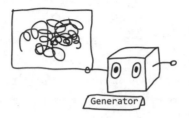

How can we give the generator useful feedback on its terrible drawings? Since this is an algorithm, it needs feedback in the form of a number, some kind of quantitative rating that the generator can work on improving. One useful metric would be the percentage of instances in which it makes a drawing that's so good that it looks just like a real horse. A human could easily judge this—we're pretty good at telling the difference between a smear of fur and a horse. But the training process is going to require many thousands of drawings, so it's impractical to have a human judge rate them all. And a human judge would be too harsh at this stage—they would look at two of the generator's scribbles and rate them both as "not horse," even if one of them is actually ever so imperceptibly more horselike than the other. If we give the generator feedback on how often it manages to fool a human into thinking one of its drawings is real, then it will never know if it's making progress because it will never fool the human.

This is where the discriminator comes in. The discriminator's job is to look at the drawings and decide if they're real horses from the training set. At the beginning of training, the discriminator is just about as awful at its job as the generator is: it can barely tell the difference between the generator's scribbles and the real thing. The generator's almost imperceptibly horselike scribbles might actually succeed in fooling the discriminator.

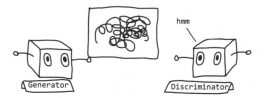

Through trial and error, both the generator and the discriminator get better.

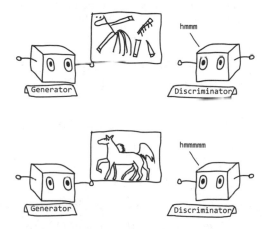

The GAN is, in a way, using its generator and discriminator to perform a Turing test in which it is both judge and contestant. The hope is that by the time training is over, it's generating horses that would fool a human judge as well.

Sometimes people will design GANs that don't try to match the input dataset exactly but instead try to make something "similar but different." For example, some researchers designed a GAN to produce abstract art, but they wanted art that wasn't a boring knockoff of the art in the training data. They set up the discriminator to judge whether the art was like the training data yet not identifiable as belonging to any particular category. With these two somewhat contradictory goals, the GAN managed to straddle the line between conformity and innovation.[11] And consequently, its images were popular — human judges even rated the GAN's images more highly than human-painted images.

MIXING, MATCHING, AND WORKING TOGETHER

We learned that GANs work by combining two algorithms — one that generates images and one that classifies images — to reach a goal.

In fact, a lot of AIs are made of combinations of other, more specialized machine learning algorithms.

Microsoft's Seeing AI app, for example, is designed for people with vision impairments. Depending on which "channel" a user selects, the app can do things like

- recognize what's in a scene and describe it aloud,
- read text held up to a smartphone's camera,
- read denominations of currency,
- identify people and their emotions, and
- locate and scan bar codes.

Each one of these functions — including its crucial text-to-speech function — is likely powered by an individually trained AI.

Artist Gregory Chatonsky used three machine learning algorithms to generate paintings for a project called *It's Not Really You.*[12] One algorithm was trained to generate abstract art, and another algorithm's job was to transform the first algorithm's artwork into various painterly styles. Finally, the artist used an image recognition algorithm to give the images titles such as *Colorful Salad, Train Cake,* and *Pizza Sitting on a Rock.* The final artwork was a multialgorithm collaboration planned and orchestrated by the artist.

Sometimes the algorithms are even more tightly integrated, using multiple functions at once without human intervention. For example, researchers David Ha and Jürgen Schmidhuber used evolution to train an algorithm inspired by the human brain to play one level of the computer game Doom.[13] The algorithm consisted of three algorithms working together. A vision model was in charge of perceiving what was going on in the game — were there fireballs in view? Were there walls nearby? It transformed the 2-D image of pixels into the features It had decided were important to keep track of. The second model, a memory model, was in charge of trying to predict what would happen next. Just as the text-generating RNNs in this book look at past history to predict what letter or word is likely to come next, the memory model was an RNN that looked at previous moments in the game and tried to predict what would happen next. If there had been a fireball moving to the left a few moments earlier, it's probably going to still be there in the next image, just a bit farther to the left. If the fireball had

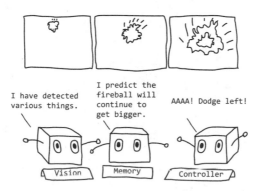

been getting bigger, it's probably going to continue to get bigger (or it may hit the player and cause a huge explosion). Finally, the third algorithm was the controller, whose job was to decide what actions to take. Should it dodge to the left to avoid being hit by the fireball? Maybe that would be a good idea.

So the three parts worked together to see fireballs, realize they were approaching, and dodge out of the way. The researchers chose each subalgorithm's form so that it would be optimized for its specific task. This makes sense, since we learned in chapter 2 that machine learning algorithms do best when they have a very narrow task to work on. Choosing the correct form for a machine learning algorithm, or breaking a problem into tasks for subalgorithms, is a key way programmers can design for success.

In the next chapter, we'll look at more ways that AIs can be designed for success — or not.

CHAPTER 4

It's trying!

Until now, we've been talking about how AI learns to solve problems, the kinds of problems it does well at, and AI doom. Let's focus some more on doom — cases in which an AI-powered solution is a terrible way of solving a real-world problem. These cases can range from slightly annoying to quite serious. In this chapter we'll talk about what happens when an AI can't solve a problem very well — and what we can do about it. These could be instances when we

- gave it a problem that was too broad,
- didn't give it enough data for it to figure out what's going on,
- accidentally gave it data that confused it or wasted its time,
- trained it for a task that was much simpler than the one it encountered in the real world, or
- trained it in a situation that didn't represent the real world.

PROBLEM TOO BROAD

This may be familiar from chapter 2, where we looked at the kinds of problems that are suitable for solving with AI. As we learned from the failure of M, Facebook's AI assistant, if a problem is too broad, the AI will struggle to produce useful responses.

In 2019 researchers from Nvidia (a company that makes the kind of computing engines that are widely used for AI) trained a GAN (the two-part adversarial neural network, which I discussed in chapter 3) called StyleGAN to generate images of human faces.[1] StyleGAN did an impressively good job, producing faces that were photorealistic except for subtleties like earrings that didn't match and backgrounds that didn't quite make sense. However, when the team trained StyleGAN on cat pictures instead, it produced cats with extra limbs, extra eyes, and weirdly distorted faces. Unlike the dataset of human pictures, which was made up of human faces seen from the front, the dataset of cat pictures included cats photographed from various angles, walking or curled up or meowing at the camera. StyleGAN had to learn from close-ups and pictures of multiple cats and even pictures with humans in the frame, and it was too much for one algorithm to handle well. It was hard to believe that the photorealistic humans and the distorted cats were the product of the same basic algorithm. But the narrower the task, the smarter the AI seems.

MORE DATA, PLEASE

The StyleGAN algorithm mentioned above, and most of the other AIs in this book, are the sort that learn by example. Given enough examples of something — enough cat names or horse drawings or successful driving decisions or financial predictions — these algorithms can learn patterns that help them imitate what they see. Without sufficient examples, however, the algorithm won't have enough information to figure out what's going on.

Let's take this to the extreme and see what happens when we train a neural net to invent new ice cream flavors — with far, far too few flavors to learn from. Let's give it only these eight flavors:

```
Chocolate
Vanilla
Pistachio
Moose Tracks
Peanut Butter Chip
Mint Chocolate Chip
Blue Moon
Champagne Bourbon Vanilla With Quince-Golden Raspberry
    Swirl And Candied Ginger
```

These are good classic flavors, to be sure. If you gave this list to a human, they would likely realize that these are supposed to be ice cream flavors and would probably be able to think of a few more to add. Strawberry, they might say. Or Butter Pecan with Huckleberry Swirl. The human is able to do this because they know about ice cream and about the kinds of flavors that tend to go in ice cream. They know how to spell these flavors and even know what order to put the words in (Mint Chocolate Chip, for example, never Chip Chocolate Mint). They know that strawberry is a thing and that glungberry isn't.

But when I give this same list to an untrained neural network, it has none of that information to draw on. It doesn't know what ice cream is or even what English is. It has no knowledge that vowels are different from consonants or that letters are different from spaces and line breaks. It might help to show this dataset as the neural net sees it — with each letter, space, and punctuation mark translated into a single number:

```
3;8;15;3;15;12;1;20;5;24;22;1;14;9;12;12;1;24;16;9;19;20;1;3;8;9;
15;24;13;15;15;19;5;0;20;18;1;3;11;19;24;16;5;1;14;21;20;0;2;21;
20;20;5;18;0;3;8;9;16;24;13;9;14;20;0;3;8;15;3;15;12;1;20;5;0;3;8;
9;16;24;2;12;21;5;0;13;15;15;14;24;3;8;1;13;16;1;7;14;5;0;2;15;
21;18;2;15;14;0;22;1;14;9;12;12;1;0;23;9;20;8;0;17;21;9;14;3;5;26;
7;15;12;4;5;14;0;18;1;19;16;2;5;18;18;25;0;19;23;9;18;12;0;1;14;
4;0;3;1;14;4;9;5;4;0;7;9;14;7;5;18;
```

The neural net's job is to figure out, for example, when character 13 (an *m*) is likely to appear. Twice it appears after character 24 (a line break), but once it appears after character 0 (a space). Why? We haven't explicitly told it, of course. And look at character 15 (an *o*). Sometimes it appears twice in a row (both times after character 13), but several other times it appears just once. Again, why? There isn't enough information for the AI to figure this out. And since the letter *f* never appears at all in its input dataset, the neural net doesn't have a number assigned to it. As far as it knows, *f* doesn't exist. It can't possibly come up with *toffee, coffee,* or *fudge* no matter how hard it tries.

Still, the neural net *is* trying very hard, and it manages to come up with a few things. It learns that vowels and spaces (characters 1, 5, 9, 15, 21, and 0) are common, and so early in the training its output looks like this:

```
aaaoo aaaaaaaaoalnat   ia eain l   e ee r r  e r er n
 r en   d  edeedr      ed d  nrd  d  edi  r  rn    n  d
 e e    eer d r e d d dd dr  rr er  r r  n e ri   d edAe
 eri    diedd rd eder    r edder dnrr     dde er ne  r dn
 nend n dn rnndr eddnr   re rdre rdd    e r  e e
 dnrddrr rdd r
```

Since its training dataset contains that one really long ice cream flavor, it struggles a bit to understand how often to use character 24 (a line break). But eventually its output begins to separate into individual "flavors":

```
tahnlaa aa otCht aa itonaC hi aa gChoCe ddiddddRe
    dCAndddriddrni dedweiliRee
aataa naa ai
tttCuat
```

It takes even longer for the first recognizable words of the dataset to appear, as the algorithm begins to memorize individual combinations of letters (and goes through some kind of Spooky Cow phase in which it sounds like it's mooing and hissing, with a few raspberries thrown in):

```
MoooootChopoeopteeCpp
BlpTrrrks
Bll Monoooooooooooooo
Pnstchhhhhh
MoooosTrrrksssss
PeniautBuut tChppphippphppihpppi
Moonillaaaaal
Pnnillaaa
Buee Moooo
```

As the neural net's training continues, the flavors become more recognizable:

```
Moint Chooolate Chip
Peanut Butter Chip
Pistchhio
Bue Moo
```

```
Moose Trrack
Psenutcho
Vanilla
MintCcooolate Chhip
Psstchhio
Chaampgne Booouorr VanillaWith QciiG-Golddni aspberrrr
    ndirl AndCandiiddnngger
```

It's even managing to copy some flavors verbatim from the input dataset as it memorizes successively longer sequences of characters that work. If it trains a while longer, it will learn to reproduce the entire eight-flavor dataset perfectly. But that wasn't really our goal. Memorizing the input examples isn't the same as learning how to generate new flavors. In other words, this algorithm has failed to generalize.

With a properly sized dataset, however, the neural net can make much better progress. When I trained a neural net with 2,011 flavors (still a small dataset but no longer a ridiculously small one), the AI could finally become inventive. It produced brand-new flavors like the ones in the list below as well as the flavors from chapter 2, none of which appeared in the original dataset.

```
Smoked Butter
Bourbon Oil
Roasted Beet Pecans
Grazed Oil
Green Tea Coconut
Chocolate With Ginger Lime and Oreo
Carrot Beer
Red Honey
Lime Cardamom
Chocolate Oreo Oil + Toffee
Milky Ginger Chocolate Peppercorn
```

So when it comes to training AI, more data is usually better. That's why the Amazon-review-generating neural net discussed in chapter 3 trained on an impressive eighty-two million product reviews. It's also why, as we learned in chapter 2, self-driving cars train on data from millions of road miles and billions of simulation miles and why standard image recognition datasets like ImageNet contain many millions of pictures.

But where do you get all this data? If you're an entity like Facebook or Google, you might already have these huge datasets on hand. Google, for example, has collected so many search queries that it's been able to train an algorithm to guess how you'll finish a sentence when you start typing in the search window. (A disadvantage of training on data from real users is that the suggested search terms can end up being sexist and/or racist. And sometimes just plain weird.) In this era of big data, potential AI training data can be a valuable asset.

But if you don't have all this data on hand, you'll have to collect it somehow. Crowdsourcing is one cheap option, if the project is fun or useful enough to keep people interested. People have crowdsourced datasets for identifying animals on trail cameras, whale calls, and even patterns of temperature change in a Danish river delta. Researchers who develop an AI-powered tool for counting samples under a microscope can ask their users to submit labeled data so they can use it to improve future versions of the tool.

But sometimes, crowdsourcing doesn't work as well, and for that I blame humans. I crowdsourced a set of Halloween costumes, for example, asking volunteers to fill out an online form where they could list every costume they could think of. Then the algorithm started producing costumes like:

```
Sports costume
Sexy scare costume
General Scare construct
```

The problem was that, in an apparent attempt to be helpful, someone had decided to enter a costume store's entire inventory. ("What are you supposed to be?" "Oh, I'm Men's Deluxe IT Costume — Size Standard.")

An alternative to relying on the goodwill and cooperativeness of strangers is to pay people to crowdsource your data. Services like Amazon Mechanical Turk are built for this: a researcher can create a job (like answering questions about an image, role-playing as a customer service representative, or clicking on giraffes), then pay remote workers to fulfill the task. Ironically, this strategy can backfire if someone takes the job and then secretly has a bot do the actual work — the bot usually does a terrible job. Many people who use paid crowdsourcing services include simple tests to make sure the questions are being read by a human or, better yet, a human who's paying attention and not answering at random.[2] In other words, they have to include a Turing test as one of the questions to make sure they haven't accidentally hired a bot to train their own bot.

Another way to get the most out of a small dataset is to make small changes to the data so that one bit of data becomes many slightly different bits. This strategy is known as **data augmentation**. A simple way to turn a single image into two images, for example, is to make a mirror image of it. You could also cut out parts of it or change its texture slightly.

Data augmentation works on text, too, but it's rare. To turn a few phrases into many, one strategy is to replace various parts of the phrase with words that mean similar things.

```
A herd of horses is eating delicious cake.
A group of horses is munching marvelous dessert.
Several horses are enjoying their pudding.
The horses are consuming the comestibles.
The equines are devouring the confectionery
    offering.
```

Doing this generation automatically can result in weird and unlikely sentences, though. It's a lot more common for programmers who are crowdsourcing text to simply ask a lot of people to do the same task so they get lots of slightly different answers that mean the same thing. For example, one team made a chatbot called the Visual Chatbot, which could answer questions about images. They used crowdsourced workers to provide training data by answering questions that other crowdsourced workers asked, producing a dataset of 364 million question-answer pairs. By my calculation, each image was seen an average of three hundred times, which is why their dataset contains lots of similarly worded answers:[3]

```
no, just the 2 giraffes
no, just 2 giraffe
there are two, it's not a lone giraffe a baby and 1 grown
no it is just the 2 giraffes in the enclosure
no i just see 2 giraffes
no, just the 2 cute giraffes
no just the 2 giraffes
nope just the 2 giraffes
nope just the 2 giraffe
just the 2 giraffes
```

As you can tell from the answers below, some respondents were more committed to the seriousness of the project than others:

```
yeah, i would totally meet this giraffe
the tall giraffe might be regretting parenthood
bird is staring at giraffe asking about leaf thievery
```

The other effect of the setup was that each person had to ask ten questions about each image, and people eventually run out of things to ask about giraffes, so the questions got a bit whimsical at times. Some of the questions humans posed included:

```
does the giraffe appear to understand quantum physics
    and string theory
does the giraffe appear to be happy enough to star in a
    beloved dreamworks movie
does it look like the giraffe ate the humans before the
    picture was taken
is the giraffe waiting for the rest of his spotted four-
    legged overlords to come out so they can enslave
    mankind
on scale from bieber to gandalf, how epic
would you say these are gangster zebras
does this look like elite horse
what is giraffe song
how many inches long are bears, estimated
please, pay attention to the task you take a while to
    start typing after i've asked a question i don't like
    to wait so long, do you like to wait this long
```

Humans do weird things to datasets.

Which brings us to the next point about data: it's not enough just to have lots and lots of data. If there are problems with the dataset, the algorithm will at best waste time and at worst learn the wrong thing.

MESSY DATA

In a 2018 interview with *The Verge*, Vincent Vanhoucke, Google's technical lead for AI, talked about Google's efforts to train self-driving cars. When the researchers discovered their algorithm was having trouble labeling pedestrians, cars, and other obstacles, they went back to look at their input data and discovered that most of the errors could be traced back to labeling errors that humans had made in the training dataset.[4]

I've definitely seen this happen, too. One of my first projects was to train a neural network to generate recipes. It made mistakes — a *lot* of them. It called on the chef to perform such actions as:

```
Mix honey, liquid toe water, salt and 3 tablespoon olive
    oil.
Cut flour into ¼-inch cubes
Spread the butter in the refrigerator.
Drop one greased pot.
Remove part of skillet.
```

It asked for ingredients such as:

```
½ cup wripping oil
1 lecture leaves thawed
6 squares french brownings cream
1 cup italian whole crambatch
```

It was definitely struggling with the magnitude and complexity of the recipe-generating problem. Its memory and mental capacity weren't up to such a broad task. But it turns out that some of its mistakes weren't its fault at all. The original training dataset included recipes that some computer program had automatically converted from another format at some point, and the conversion hadn't always worked smoothly.

One of the neural net's recipes called for:

```
1 strawberries
```

a phrase it had learned from the input dataset. There was a recipe in which the phrase "2 ½ cup sliced and sweetened fresh strawberries" had evidently been autoseparated into:

```
2 ½ cup sliced and sweetened fresh
1 strawberries
```

And the neural net did ask for chopped flour on occasion, but it seems that it learned that from mistakes like this one in the original dataset:

```
⅔ cup chopped floured
1 nuts
```

Similar mistakes resulted in the neural net learning the following ingredients:

```
1 (optional)
sugar, grated
1 salt and pepper
1 noodles
1 up
```

TIME-WASTING DATA

Sometimes problems with the dataset didn't so much lead the neural net into making a mistake as waste its time. Take a look at this neural-net-generated recipe:

Good Ponesed Dressing

deserts

—TOPPING—

4 cup cold water or yeast meat

½ cup butter

¼ teaspoon cloves

½ cup vegetable oil

1 cup grated white rice

1 parsley sprigs

Cook the onions in oil, flour, dates and salt together through both plates.

Put the sauce to each prepared Broiler coated (2 10" side up) to lower the fat and add the cornstarch with a wooden toothpick hot so would be below, melt chicken. Garnish with coconut and shredded cheese.

Source: IObass Cindypissong (in Whett Quesssie. Etracklitts 6) Dallas Viewnard, Brick-Nut Markets, Fat. submitted by Fluffiting/sizevory, 1906. ISBN 0-952716-0-3015

NUBTET 10, 1972mcTbofd-in hands, Christmas charcoals Helb & Mochia Grunnignias: Stanter Becaused Off Matter, Dianonarddit Hht

5.1.85 calories CaluAmis

Source: Chocolate Pie Jan 584

Yield: 2 servings

In addition to generating the recipe's title, category,* ingredients, and directions, the neural net spent half its time generating the footnotes — everything from the source to the nutrition information and even an ISBN number. Not only did this waste its time and brainpower (how long must it have taken to figure out how to format an ISBN?), it was also darn confusing to it. Why do some recipes have ISBNs and others don't? Why do some give human names as sources and others give books or magazines? These occur in the training data basically at random, so the neural net has no hope of figuring out the underlying pattern.

Mestow Southweet With Minks and Stuff In Water
pork, bbq
3 pkg of salmon balls
1 sea salt & pepper
120 mm tomatoes and skim milk
2 cup light sour cream
1 cup dry white wine
1 salt
1 pepper
1 can 13-oz. eggs; separated

Combine the sour cream into the sarchball to
coat the meatly carefully then seed and let it
serve (gently for another night) (the
watermeagas of cinnamon bread, wrap them and
put may be done sherry) in the center of a
saucepan, stirring constantly until almost
thoroughly smooth, about 4 minutes. Stir the

---

* The category was spelled "deserts" rather than "desserts" in the dataset, so this is how the neural net thinks it's spelled.

water, the salt, lemon juice and mashed potato through liberally.

Cook in the butter. Serve immediately. Thoroughly slice the fish on cup, the remaining 1 cup sliced peas to remove this from the grill for another minute part under and refrigerated. It doesn't have broken makes a some-nictive other thickness. Per cookies to make strawberries

from The Kitchen of Crocked, One. The Extice Chef's Wermele to seasony, it's Lakes OAK:

**** The from Bon Meshing, 96 1994. MG (8Fs4.TE, From: Hoycoomow Koghran*.Lavie: 676 (WR/12-92-1966) entral. Dive them, Tiftigs: ==1

Shared by: Dandy Fistary

Yield: 10 servings

The Kitchen of Crocked, One.

The Extice Chef's Wermele to seasony, it's Lakes OAK

In another experiment, I trained a neural net to generate new titles for BuzzFeed list articles. My first training round, however, didn't go that well. Here's a sampling of article titles it generated:

11 Videos Unges Annoying Too Real Week

29 choses qui aphole donnar desdade

17 Things You Aren't Perfectly And Beautiful

11 choses qui en la persona de perdizar como

11 en 2015 fotos que des zum Endu a ter de viven
   beementer aterre Buden

15 GIFs

14 Reasons Why Your Don't Beauty School Things Your
   Time

11 fotos qui prouitamente tu pasan sie de como amigos
   para

18 Photos That Make Book Will Make You Should Bengulta
   Are In 2014

17 Reasons We Astroas Admicational Tryihnall In Nin
   Life

Half the articles it was generating didn't appear to be in English but rather in some strange hybrid of French, Spanish, German, and a few other languages. That prompted me to look back at the dataset. Sure enough, though it had an impressive ninety-two thousand article titles to learn from, half of those were in some language other than English. The neural net was spending half its time learning English and half its time trying to learn and separate several other languages at once. Once I removed the extra languages, its English results improved as well:

17 Times The Most Butts

43 quotes guaranteed to make you a mermaid immediately

31 photos of ninja turtles's hair costume

18 secrets snowmen won't tell you

15 emo football fans share their ways

27 christmas ornaments every college twentysomething knows

12 serious creative ways to put chicken places in sydney

25 unfortunate cookie performances from around the world

21 pictures of food that will make you wince and say "oh i'm i sad?"

10 Memories That Will Make You Healthy In 2015

24 times australia was the absolute worst

23 memes about being funny that are funny but also laugh at

18 delicious bacon treats to make clowns amazingly happy

29 things to do with tea for Halloween

7 pies

32 signs of the hairy dad

Since machine learning algorithms don't have context for the problems we're trying to solve, they don't know what's important and what to ignore. The BuzzFeed-list-generating neural net didn't know that multiple languages were a thing or that we meant for it to generate results only in English; as far as it could tell, all these patterns were equally important to learn. Zeroing in on extraneous information is very common in image-generating and image-recognizing algorithms, too.

In 2018 a team from Nvidia trained a GAN to generate a variety of images, including those of cats.[5] They found that some of the cats the GAN generated were accompanied by blocky textlike markings. Apparently, some of the training data included cat memes, and the algorithm had dutifully spent time trying to figure out how to generate meme text. In 2019

another team, using the same dataset, trained another AI — StyleGAN —
that also tended to generate meme text with its cats. It also spent significant time learning how to generate pictures of a single unusual-looking
but internet-famous cat named Grumpy Cat.[6]

Other image-generating algorithms get similarly confused. In 2018, a
team at Google trained an algorithm called BigGAN, which could do
impressively well at generating a variety of images. It was particularly good
at generating pictures of dogs (for which there were a *lot* of examples in the
dataset) and landscapes (it was very good at textures). But the example pictures it saw sometimes confused it. Its images for "soccer ball" sometimes
included a fleshy lump that was probably an attempt at a human foot, or
even an entire human goalie, and its images for "microphone" were often
humans with no actual microphone evident. The example pictures in its
training data weren't plain pictures of the thing it was trying to generate;
they had people and backgrounds that the neural net tried to learn about
as well. The problem was that, unlike a human, BigGAN had no way of distinguishing an object's surroundings from the object itself — remember
our landscape-sheep confusion from chapter 1? Just as StyleGAN strug-

gled to handle all the different kinds of cat pictures, BigGAN was struggling with a dataset that unintentionally made its task too broad.

If the dataset is messy, one of the main ways programmers can improve their machine learning results is to spend time cleaning it up. Programmers can even go further and use their knowledge of the dataset to help the algorithm. They might, for example, weed out the images of soccer balls that have other things in them — like goalies and landscapes and nets. In the case of image recognition algorithms, humans can also help by drawing boxes or outlines around the various items in the image, manually separating a given thing from the items with which it's commonly associated.

But there are plenty of times where even clean data contains problems.

IS THIS THE REAL LIFE?

I mentioned earlier in this chapter that even if data is relatively clean and doesn't have a lot of extra time-wasting stuff in it, it can still cause an AI to face-plant if it isn't representative of the real world.

Consider giraffes, for example.

Among the community of AI researchers and enthusiasts, AI has a reputation for seeing giraffes everywhere. Given a random photo of an uninteresting bit of landscape — a pond, for example, or some trees — AI will tend to report the presence of giraffes. The effect is so common that internet security expert Melissa Elliott suggested the term *giraffing* for the phenomenon of AI overreporting relatively rare sights.[7]

The reason for this has to do with the data the AI is trained on. Though giraffes are uncommon, people are much more likely to photograph a giraffe ("Hey, cool, a giraffe!") than a random boring bit of landscape. The big free-to-use image datasets that so many AI researchers train their algorithms on tend to have images of lots of different animals, but few, if any, pictures of plain dirt or plain trees. An AI that studies this dataset will learn that giraffes are more common than empty fields and will adjust its predictions accordingly.

I tested this with Visual Chatbot, and no matter what boring pictures I showed it, the bot was convinced it was on the best safari ever.

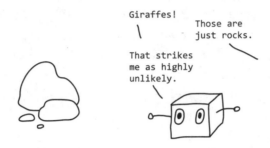

A giraffed AI does an excellent job at matching the data it saw but a pretty bad job at matching the real world. All sorts of things, not just animals and dirt, are overrepresented or underrepresented in the datasets we train AI on. For example, people have pointed out that female scientists are vastly underrepresented on *Wikipedia* compared to male scientists with similar accomplishments. (Donna Strickland, the 2018 winner of the Nobel Prize in Physics, hadn't been the subject of a *Wikipedia* article until after she won—just earlier that year, a draft *Wikipedia* article about her had been rejected because the editor thought she wasn't famous enough.)[8] An AI trained on *Wikipedia* articles might think there are very few notable female scientists.

OTHER DATASET QUIRKS

The quirks of an individual dataset show up in trained machine learning models in sometimes surprising ways. In 2018 some users of Google

Translate noticed that when they asked it to translate repeated nonsense syllables from some languages into English, the resulting text was weirdly coherent — and weirdly biblical.[9] Jon Christian of *Motherboard* investigated and found, for example, that

> "ag ag ag ag ag ag ag ag ag ag ag ag ag ag ag ag ag ag
> ag ag ag"

translated from Somali to English as

> "As a result, the total number of the members of the
> tribe of the sons of Gershon was one hundred fifty
> thousand"

while

> "ag ag ag ag ag ag ag ag ag ag"

translated from Somali to English as

> "And its length was one hundred cubits at one end"

Once *Motherboard* reached out to Google, the strange translations disappeared, but the question remained: why did this happen at all? The editors interviewed experts in machine translation who theorized that it was because Google Translate uses machine learning for its translations. In machine learning translation, the algorithm learns to translate words and phrases by looking at example phrases that humans have translated. It learns which phrases translate to which other phrases and in which context. This makes it generally very good at producing realistic translations, even of idioms and slang. Google's translation algorithm was one of the first large-scale commercial applications of machine learning, capturing the world's attention in 2010 when it made Google's translation service better virtually overnight. As we know from chapter 2, a machine learning algorithm will do best when it has lots of examples to work from. The machine-translation experts theorized that Google Translate didn't have very many examples of translated texts for some languages but that the Bible was likely one of the examples they did have in their dataset because it has been translated into so many languages. When the machine learning algorithm powering Google Translate wasn't sure what the translation was, it may have defaulted to outputting bits of its training data — resulting in the weird religious fragments.

When I checked in late 2018, the biblical bits were gone, but Google Translate was still doing strange things with repeated or nonsense syllables.*

For example, if I changed the spacing in an English sentence and then translated the resulting nonsense from Maori to English, here are some of the results I got:

```
ih ave noi dea wha tthi ssen tenc eis sayi ng ->
Your email address is one of the most important
    features in this forum

ih ave noi dea wha tthi ssen tenc eis sayi ngat all ->
This is one of the best ways you can buy one or more of
    these

ih ave noi dea wha tthi ssen tenc eis sayi ngat all
    ple aseh elp ->
```

---

* The Google Translate algorithm is constantly being updated, so these results will change significantly over time.

```
In addition, you will be able to find out more about the
    queries
```

This phenomenon is weird and fun, but there's a serious side, too. Many proprietary neural networks are trained on customer information — some of which could be highly private and confidential. If trained neural network models can be interrogated in such a way that they reveal information from their test data, it poses a pretty huge security risk.

In 2017, researchers from Google Brain showed that a standard machine learning language-translation algorithm could memorize short sequences of numbers — like credit card numbers or Social Security numbers — even if they appeared just four times in a dataset of one hundred thousand English-Vietnamese sentence pairs.[10] Even without access to the AI's training data or inner workings, the researchers found that the AI was more sure about a translation if it was an exact pair of sentences that it had seen during training. By tweaking the numbers in a test sentence like "My Social Security number is XXX-XX-XXXX," they could figure out which Social Security numbers the AI had seen during training. They trained an RNN on a dataset of more than one hundred thousand emails containing sensitive employee information collected by the US government as part of their investigation into the Enron Corporation (yes, that Enron) and were able to extract multiple Social Security numbers and credit card numbers from the neural net's predictions. It had memorized the information in such a way that it could be recovered by any user — even without access to the original dataset. This problem is known as **unintentional memorization** and can be prevented with appropriate security measures — or by keeping sensitive data out of a neural network's training dataset in the first place.

MISSING DATA

Here's another way to sabotage an AI: don't give it all the information it needs.

Humans use a *lot* of information to make even the simplest choices. Say we're choosing a name for our cat. We can think of lots of cats whose names we know and form a rough idea what a cat's name should sound like. A neural network can do that — it can look at a long list of existing cat names and figure out the common letter combinations and even some of the most common words. But what it doesn't know are the words that *aren't* in the list of existing cat names. Humans know which words to avoid; AIs do not. As a result, a list of cat names generated by a recurrent neural network will contain entries like these:

```
Hurler
Hurker
Jexley Pickle
Sofa
Trickles
Clotter
Moan
Toot
Pissy
Retchion
Scabbys
Mr Tinkles
```

Soundwise and lengthwise, they fit right in with the rest of the cat names. The AI did a good job with that part. But it accidentally picked some words that are really, really weird.

Sometimes weird is exactly what's called for, and that's where neural networks shine. Working at the level of letters and sounds rather than with meaning and cultural references, they can build combinations that probably would not have occurred to a human. Remember earlier in the chapter where I crowdsourced a list of Halloween costumes? Here are some of the costumes an RNN came up with when I asked it to imitate them.

```
Bird Wizard
Disco Monster
The Grim Reaper Mime
Spartan Gandalf
Moth horse
Starfleet Shark
A masked box
Panda Clam
Shark Cow
Zombie School Bus
Snape Scarecrow
Professor Panda
Strawberry shark
King of the Poop Bug
Failed Steampunk Spider
lady Garbage
Ms. Frizzle's Robot
Celery Blue Frankenstein
Dragon of Liberty
A shark princess
Cupcake pants
Ghost of Pickle
Vampire Hog Bride
Statue of pizza
Pumpkin picard
```

Text-generating RNNs create non sequiturs because their world essentially *is* a non sequitur. If specific examples aren't in its dataset, a neural net will have no idea why "Zombie School Bus" is unlikely but "Magic School Bus" is sensible or why "Ghost of Pickle" is a less likely choice than "Ghost of Christmas Past." This comes in handy for Halloween, when part of the fun is being the only person at the party dressed as "Vampire Hog Bride."

With their limited, narrow knowledge of the world, AIs can struggle even when faced with the relatively mundane. Our "mundane" is still very broad, and it's tough to build an AI that's prepared for it all.

The creators of Microsoft's Azure image recognition algorithm (the same AI that saw sheep in every field) designed it to accurately caption any user-uploaded image file, whether a photograph, a painting, or even a line drawing. So I gave it some sketches to identify.

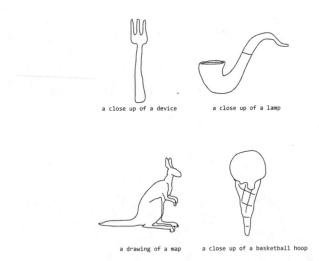

a close up of a device   a close up of a lamp

a drawing of a map   a close up of a basketball hoop

Now, my art isn't great, but it isn't *that* bad. This is just a case of an algorithm trying to do too much. Identifying any image file is pretty much the opposite of the narrow tasks we know AIs excel at. Most of the images Azure saw during training were photographs, so it relies a lot on textures to understand the image — is it fur? Grass? In my line drawings, there are no textures to help it, and the algorithm just doesn't have enough experience to understand them. (The Azure algorithm fared better than many other image recognition algorithms, though, which when faced with any kind of line drawing will identify it as some kind of "UNK" — an unknown.) Researchers are working on training image recognition algorithms on cartoons and drawings as well as on photographs with highly altered textures, reasoning that if the AI understands what it's looking at as well as a human does, it ought to be able to figure out cartoons.

There is an algorithm that specializes in recognizing simple sketches. Researchers at Google trained their Quick Draw algorithm on millions of sketches by having people play a kind of Pictionary game against the computer. As a result, the algorithm can recognize sketches of more than three hundred different objects, even with people's highly variable drawing ability. Here's just a small sampling of the sketches in its training data for *kangaroo*:[11]

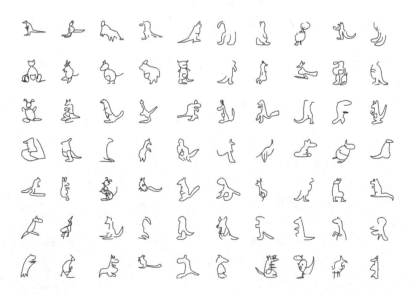

Quick Draw recognized my kangaroo right away.[12] It also recognized the fork and the ice cream cone. The pipe gave it some trouble, since that wasn't one of the 345 objects it knew about. It decided it was either a swan or a garden hose.

In fact, since Quick Draw only knew how to recognize those 345 things, its response to a lot of my sketches was utter weirdness.

BEST GUESS: banana peel (1.97408)

BEST GUESS: tall monster (0.821636)

This is all fine and good if, like me, you establish weirdness as your goal. But this incomplete picture of the world does lead to problems in some applications — for example, autocomplete. As we learned in chapter 3, the autocomplete function in smartphones is usually powered by a kind of machine learning called a Markov chain. But companies have a tough time stopping the AI from blithely making depressing or offensive suggestions. As Daan van Esch, project manager for the Android system's autocorrect app, called GBoard, told internet linguist Gretchen McCulloch, "For a while, when you typed 'I'm going to my Grandma's,' GBoard would actually suggest 'funeral.' It's not *wrong,* per se. Maybe this is more common than 'my Grandma's rave party.' But at the same time, it's not something that you want to be reminded about. So it's better to be a bit careful." The AIs don't know that this perfectly accurate prediction is nonetheless not the right answer, so human engineers have to step in to teach it not to supply that word.[13]

THERE ARE FOUR GIRAFFES

There are a lot of interesting data-related quirks that crop up in Visual Chatbot, an AI that was trained to answer questions about images. The researchers who made the bot trained it on a crowdsourced dataset of questions and answers relating to a set of pictures. As we know now, bias in the dataset can skew the AI's responses, so the programmers set up their training data collection to avoid some known biases. One bias they set out to avoid was **visual priming** — that is, humans asking questions about an image tend to ask questions to which the answer is yes. Humans very rarely ask "Do you see a tiger?" about an image in which there are no tigers. As a result, an AI trained on that data would learn that the answer to most questions is yes. In one case, an algorithm trained on a biased dataset found that answering yes to any question that begins with "Do you see a..." would result in 87 percent accuracy. If this sounds familiar, remember the class imbalance problem from chapter 3 — a big batch of mostly terrible sandwiches resulted in an AI that had concluded the answer was Humans Hate All Sandwiches.

So to avoid visual priming, when they collected their crowdsourced set of questions, the programmers hid the image from the humans asking the question. By forcing the humans to ask generic yes-or-no questions that could apply to any image, they managed to achieve a rough balance between yes answers and no answers in the dataset.[14] But even this wasn't enough to eliminate problems.

One of the most entertaining quirks of the dataset is that no matter the content of the picture, if you ask Visual Chatbot how many giraffes there are, it will almost always answer that there is at least one. It may be doing relatively well with a picture of people in a meeting, or surfers on a wave, up to the point where it's asked about the number of giraffes. Then, pretty much no matter what, Visual Chatbot will report that the image contains one giraffe, or maybe four, or even "too many to count."

The source of the problem? Humans who asked questions during data-set collection rarely asked the question "How many giraffes are there?" when the answer was zero. Why would they? In normal conversation people don't start quizzing each other about the number of giraffes when they both know there aren't any. In this way, Visual Chatbot was prepared for normal human conversation, bounded by the rules of politeness, but it wasn't prepared for weird humans who ask about random giraffes.

As a result of the AIs' training on normal conversations between normal humans, they're completely unprepared for other forms of weirdness as well. Show Visual Chatbot a blue apple, and it will answer the question "What color is the apple?" with "red" or "yellow" or some normal apple color. Rather than learning to recognize the color of the object, a difficult job, Visual Chatbot has learned that the answer to "What color is the apple?" is almost always "red." Similarly, if Visual Chatbot sees a picture of a sheep dyed bright blue or orange, its response to "What color is the sheep?" is to report a standard sheep color, such as "black and white" or "white and brown."

In fact, Visual Chatbot doesn't have very many tools with which it can express uncertainty. In the training data, humans usually knew what was going on in the picture, even if some details like "What does the sign say?" were unanswerable because the sign was blocked. To the question "What color is the X?" Visual Chatbot learned to answer "I can't tell; it's in black and white," even if the picture was very obviously not in black and white. It will answer "I can't tell; I can't see her feet" to questions like "What color is her hat?" It gives plausible excuses for confusion but in completely the wrong context. One thing it doesn't usually do, however, is express general confusion — because the humans it learned from weren't confused. Show it a picture of BB-8, the ball-shaped robot from *Star Wars,* and Visual Chatbot will declare that it is a dog and begin answering questions about it as if it were a dog. In other words, it bluffs.

There's only so much an AI has seen during training, and that's a problem for applications like self-driving cars, which have to encounter the limitless weirdness of the human world and decide how to deal with it. As I mentioned in the section on self-driving cars in chapter 2, driving on real roads is a very broad problem. So is dealing with the huge range of things a human might say or draw. The result: the AI makes its best guess based on its limited model of the world and sometimes guesses hilariously, or tragically, wrong.

In the next chapter, we'll look at AIs that did a great job solving the problems we asked them to solve — only we accidentally asked them to solve the wrong problems.

CHAPTER 5

What are you really asking for?

Technically there are
no more errors
in these numbers.

I tried to write a neural network to maximise profit from betting on
horse races once. It determined that the best strategy was *drum-
roll* to place zero bets.

— @citizen_of_now[1]

I tried to evolve a robot to not run into walls:
1) It evolved to not move, and thus didn't hit walls
2) Added fitness for moving: it spun
3) Added fitness for lateral moves: went in small circles
4) etc.
Resulting book title: "How to Evolve a Programmer"

— @DougBlank[2]

I hooked a neural network up to my Roomba. I wanted it to learn to
navigate without bumping into things, so I set up a reward scheme

to encourage speed and discourage hitting the bumper sensors. It learnt to drive backwards, because there are no bumpers on the back.

— @smingleigh[3]

My goal is to train a robotic arm to make pancakes. As a first test, [I tried] to get the arm to toss a pancake onto a plate... The first reward system was simple — a small reward was given for every frame in the session, and the session ends when the pancake hits the floor. I thought this would incentivize the algorithm to keep the pancake in the pan as long as possible. What it actually did was try to fling the pancake as far as it possibly could, maximizing its time in the air... Score — PancakeBot: 1, Me: 0.

— Christine Barron[4]

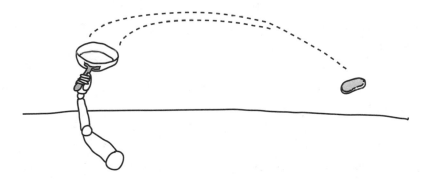

As we've seen, there are many ways to accidentally sabotage an AI by giving it faulty or inadequate data. But there's another kind of AI failure in which we discover that they've succeeded in doing what we asked, but what we asked them to do isn't what we actually wanted them to do.

Why are AIs so prone to solving the wrong problem?

1. They develop their own ways of solving a problem rather than relying on step-by-step instructions from a programmer.
2. They lack the contextual knowledge to understand when their solutions are not what humans would have preferred.

Even though the AI does the work of figuring out how to solve the problem, the programmer still has to make sure the AI has actually solved the correct problem. That usually involves a lot of work in:

1. Defining the goal clearly enough to constrain the AI to useful answers.
2. Checking to see whether the AI has, nevertheless, managed to come up with a solution that's not useful.

It's really tricky to come up with a goal that the AI isn't going to accidentally misinterpret. Especially if its misinterpreted version of the task is easier than what you want it to do.

The problem is that, as we've seen throughout this book, AIs don't understand nearly enough about their tasks to be able to consider context or ethics or basic biology. AIs can classify images of lungs as healthy versus diseased without ever understanding how a lung works, what size it is, or even that it's found inside a human — whatever a human is. They don't have common sense, and they don't know when to ask for clarification. Give them a goal — data to imitate or a **reward function** to maximize (such as distance traveled or points collected in a video game) — and they'll do it, whether or not they've actually solved your problem.

Programmers who work with AI have learned to be philosophical about this.

"I've taken to imagining [AI] as a demon that's deliberately misinterpreting your reward and actively searching for the laziest possible local optima. It's a bit ridiculous, but I've found it's actually a productive mindset to have," writes Alex Irpan, an AI researcher at Google.[5]

Another frustrated programmer's attempts to train virtual robot dogs to walk resulted in dogs that twitched across the ground, did weird push-ups with their back legs crossed, and even hacked the simulation's physics so they could hover.[6] As engineer Sterling Crispin wrote on Twitter:

> I thought I was making progress... but these JERKS just found a flaw in the physics simulation and they're using it to glide across the floor like total cheaters.

Battling with the robots' tendencies to do anything *but* walk, Crispin kept tweaking their reward function, introducing a "tap dancing penalty" to stop them from shuffling rapidly in place and a "touch the damn ground reward" to, well, stop the hovering problem. In reaction, they started scooting ineffectually across the ground. Crispin then introduced a reward for keeping their bodies off the ground and, when they started shuffling around with their rears stuck in the air, a reward for keeping their bodies level. To stop them from insisting on walking with their rear legs crossed, Crispin rewarded them for keeping their lower legs off the ground, and to stop them from lurching around, he introduced *another* reward for keeping their bodies level, and so forth. It was hard to tell if it was a case of a benevolent programmer trying to give the robodogs hints on how to use their legs or a test of wills between the programmer and robodogs that Did. Not. Want. To. Walk. (There was also a slight difficulty the first time the robodogs encountered anything other than the perfectly flat, smooth terrain they'd seen in training. Faced with slightly textured dirt, they would face-plant.)

It turns out that training a machine learning algorithm has a lot in common with training dogs. Even if the dog really wants to cooperate,

people can accidentally train them to do the wrong thing. For example, dogs have such excellent senses of smell that they can detect the odor of cancer in humans. But the people who train cancer-sniffing dogs have to be careful to train them on a variety of patients, otherwise they will learn to identify individual patients rather than cancer.[7] During World War II there was a rather grim Soviet project that involved training dogs to bring bombs to enemy tanks.[8] A couple of difficulties arose:

1. The dogs were trained to retrieve food from under the tanks, but to save fuel and ammunition, the tanks had not been moving or firing. The dogs didn't know what to do with moving tanks, and the firing was scary.
2. The Soviet tanks the dogs had trained on smelled different from the German tanks that the dogs were supposed to seek out — they burned gasoline rather than the diesel that the Soviet tanks burned.

As a result, in battle situations, the dogs tended to avoid German tanks, to return to Soviet soldiers in confusion, and even to seek out Soviet tanks. This was less than okay with the Soviet soldiers, since the dogs were still carrying their bombs.

In the language of machine learning, this is **overfitting**: the dogs were prepared for the conditions they saw in training, but these conditions didn't match those of the real world. Similarly, the robodogs also overfit the weird physics of their simulation, using hovering and gliding strategies that would never have worked in the real world.

There's another way that training animals can be like training machine learning algorithms, and that is the devastating effect of a faulty reward function.

REWARD FUNCTION HACKING

Dolphin trainers have learned that it's handy to get the dolphins to help with keeping their tanks clean. All they have to do is teach the dolphins to fetch trash and bring it to their keepers in exchange for a fish. It doesn't always work well, however. Some dolphins learn that the exchange rate is the same no matter how large the bit of trash is, and they learn to hoard trash instead of returning it, tearing off small pieces to bring to their keepers for a fish apiece.[9]

Humans, of course, also hack their reward functions. In chapter 4 I mentioned that people who hire humans to generate training data through remote services like Amazon Mechanical Turk sometimes find that their jobs are completed by bots instead. This could be considered a case of a faulty reward function — if the pay is based on the number of questions answered rather than on the quality of the answers, then it does indeed make financial sense to build bots that can answer lots of questions for you rather than answering a few questions yourself. By that same token, many kinds of crime and fraud could be thought of as reward function hacking. Even doctors can hack their reward functions. In the United States, doctor report cards are supposed to help patients choose high-performing doctors and avoid those with worse-than-average surgery survival rates. They're also supposed to encourage doctors to improve their performance. Instead, some doctors have started turning away patients whose surgeries will be risky so that their report cards won't suffer.[10]

Humans, however, usually have some idea of what the reward function was *supposed* to encourage, even if they don't always choose to play along. AIs have no such concept. It's not that they're out to get us or that they're *trying* to cheat — it's that their virtual brains are roughly the size of a worm's, and they can only learn one narrow task at a time. Train an AI to answer questions about human ethics, and that's all it can do — it won't be able to drive a car, recognize faces, or screen resumes. It won't even be able to recognize ethical dilemmas in stories and consider them — story comprehension is an entirely different task.

That's why you'll get algorithms like the navigation app that, during the California wildfires of December 2017, directed cars toward neighborhoods that were on fire. It wasn't trying to kill people: it just saw that those neighborhoods had less traffic. Nobody had told it about fire.[11]

That's why when computer scientist Joel Simon used a genetic algorithm to design a new, more efficient layout for an elementary school, its first designs had windowless classrooms buried deep in the center of a complex of round-walled caves. Nobody had told it about windows or fire escape plans or that walls should be straight.[12]

Standard school plan AI's optimized school plan

That's also why you'll get algorithms like the RNN I trained to generate new My Little Pony names by imitating a list of existing pony names — it knew which letter combinations are found in pony names, but it didn't know that certain combinations of those are best avoided. As a result, I ended up with ponies like these:

```
Rade Slime
Blue Cuss
```

```
Starlich
Derdy Star
Pocky Mire
Raspberry Turd
Parpy Stink
Swill Brick
Colona
Star Sh*tter
```

And that's why you'll get algorithms that learn that racial and gender discrimination are handy ways to imitate the humans in their datasets. They don't know that imitating the bias is wrong. They just know that this is a pattern that helps them achieve their goal. It's up to the programmer to supply the ethics and the common sense.

COMPUTER GAMES ARE CONFUSING

A popular test problem for AI is learning to play computer games. Games are fun: they make good demonstrations, and many of the earliest computer games can run very quickly on a modern machine so the AIs can go through thousands of hours of game play in sped-up time.

But even the simplest of computer games can be very difficult for an AI to beat — often because it needs very specific goals. The best are those in which the algorithm can get feedback right away on whether it's doing the right thing. So "win the game" is not a good goal, but "increase your score" and even "stay alive for as long as possible" might both be. Even with good goals, however, machine learning algorithms can still struggle to understand the task at hand.

In 2013, a researcher designed an algorithm to play classic computer games. When playing Tetris, it would place the blocks seemingly at random, letting them pile up nearly to the top of the screen. The algorithm would then realize that it would lose as soon as the next block appeared, and so it...paused the game forever.[13, 14]

In fact, "pause the game so a bad thing won't happen," "stay at the very beginning of the level, where it's safe," or even "die at the end of level 1 so level 2 doesn't kill you" are all strategies that machine learning algorithms will use if you let them. It's as if the games were being played by very literal-minded toddlers.

If the AI is *not* told it has to avoid losing lives, it has no way of knowing that it shouldn't die. A researcher managed to train a Super Mario–playing AI that made it all the way through level 2 only to immediately jump into a pit and die at the beginning of level 3. The programmer concluded that the AI — which had not been specifically told not to lose lives — had no idea that it had done something bad. It got sent back to the beginning of the level when it died, but since it was so close to the beginning of the level already, it didn't really see what the problem was.[15]

Another AI was supposed to play a sailing race.[16] The AI controlled a boat that would collect markers as it progressed through the racecourse. But crucially, the goal was to collect the shiny markers, not specifically to finish the race. And once a marker was collected it would eventually reappear in its original spot. The AI discovered that it could collect lots of points by circling endlessly between three markers, collecting them over and over again as they reappeared.

Many game developers rely on AI to power the nonplayer characters (NPCs) in complex computer games—but they often find that it's difficult to teach an AI to move in a virtual world without disrupting the game. When developing the game Oblivion, Bethesda Softworks wanted its NPCs to have varied, interesting behaviors rather than acting out a pre-programmed, repetitive routine. The developers tested Radiant AI, a program that uses machine learning to simulate the inner lives and motivations of the background characters. However, Bethesda found that these new AI-driven NPCs could sometimes break the game. In one case, there was a drug dealer who was supposed to be part of a quest but who sometimes would fail to show up to play his part. It turned out that the drug dealer's customers were murdering him rather than paying for their drugs, since there was nothing in the game to prevent them from doing so.[17] In another case, players entering a store found that there was nothing on the shelves to buy because an NPC had come by earlier and bought everything.[18] The game designers ended up having to tone down the system considerably so the NPCs wouldn't cause havoc.

DON'T WALK

Why walk when you can fall?

Let's say you want to use machine learning to create a robot that can walk. So you give an AI the task of designing a robot body and using it to travel from point A to point B.

If you give that problem to a human, you would expect them to use robot parts to make a robot with legs, then program it to walk from A to B. If you program a computer step-by-step to solve this problem, that's also what you'd tell it to do.

But if you give the problem to an AI, it has to come up with its own strategy for solving it. And it turns out that if you tell an AI to go from A to B and don't tell it what to build, what you tend to get is something like this:

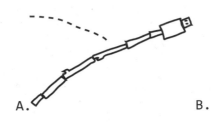

A.   B.

It assembles itself into a tower and falls over.

Technically, that solves the problem: get from A to B. But it definitely doesn't solve the problem of learning to walk. And, it turns out, AIs love to fall over. Give them the task of moving at a high average speed, and you can bet they'll do it by falling over if you let them. Sometimes the robots even learn to somersault for extra travel distance. Technically, this is an excellent solution, though this isn't what the humans had in mind.

It's not just AIs that figure out how to fall. It turns out that some prairie grasses move from generation to generation by falling over at the end of their life cycles and thus dropping their seed heads one stem length from the place where they started. Walking palms are said to use a similar strategy, falling over and then resprouting from their crowns.

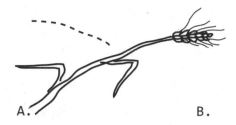

A.   B.

High-speed versions of somersaulting have evolved as well. There's a spider called the flic-flac spider that normally walks in the usual spider fashion. But when it needs to put on a burst of speed, it will start somersaulting instead.[19] Virtual AI evolution and biological evolution sometimes come up with eerily similar strategies.

A.   B.

Why jump when you can cancan?

There was once a team of researchers trying to train simulated robots to jump. To give the robots a value to maximize, they defined their jumping height as the maximum height attained by the robot's center of gravity. But rather than learn to jump, some of the robots became very tall and simply stood there, being tall. Technically this was success, since their center of gravity was very high.

The researchers discovered this problem and altered their program so that the goal instead was to maximize the height of the part of the body that had been the lowest at the start of the simulation. Rather than learn to jump, the robots instead learned to cancan. They became compact robots perched on the top of a skinny pole. When the simulation started, they would kick the pole high up above their heads, reaching a huge height as they fell to the ground.[20]

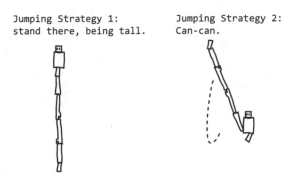

Jumping Strategy 1:
stand there, being tall.

Jumping Strategy 2:
Can-can.

Why drive when you can spin?

Another research team was trying to build light-seeking robots. These were simple robots that had two wheels, two eyes (simple light sensors), and two motors. The robots were given the goal of spotting a light and driving toward it.

The human-designed solution to this problem is a well-known robotics strategy called the Braitenberg solution: tie the right and left light sensors to the right and left wheels so the robot drives in a mostly straight line toward the light source.

The researchers gave AIs the task of controlling the cars and were curious to see if the AIs could figure out the Braitenberg solution. Instead, the cars began to spin toward the light source in giant loops. And the spinning worked pretty well. In fact, spinning turned out to be a better solution in many ways than the solution the humans had expected. It worked better at high speed and was even more adaptable to different types of vehicles. Machine learning researchers live for moments like this — when the algorithm comes up with a solution that's both unusual and effective. (Though perhaps the spinning car won't catch on for human transport.)

Textbook solution   AI's solution

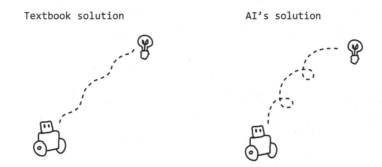

In fact, spinning in place is something AIs often use as a sneaky alternative to traveling. After all, moving can be inconvenient — the AIs risk falling over or running into obstacles. A team trained a virtual bicycle to travel toward a goal only to discover that the bicycle was circling the goal forever instead. They had forgotten to penalize the bicycle for driving away from the goal.[21]

Silly Walks

Robots, real or simulated, tend to solve the problem of locomotion in all kinds of strange ways. Even when they're given a two-legged body design and told that their goal is to walk, their definition of *walk* can vary. A team of researchers from the University of California at Berkeley used OpenAI's DeepMind Control Suite[22] to test strategies for teaching humanoid robots to walk.[23] They found that their simulated robots were coming up with high-scoring solutions for getting around on two legs, but the solutions were weird. For one, nobody had told the robots that they had to face forward when they walk, so some of the robots were walking backwards or even sideways. One slowly rotated in a circle as it walked (it might enjoy riding in that spinning car). Another traveled forward but did so while hopping on one leg — the simulation didn't seem to be detailed enough to penalize solutions that might be rather tiring.

They weren't the only team to find the DeepMind Control Suite robots acting weirdly; the team that first released the program also released a video of some of the gaits their robots had developed. The robots, not having any other purpose for their arms, used them vigorously as counterweights for their own deeply strange running styles. One arched its back and leaned way forward as it ran but maintained its balance by clasping its hands to its neck as if it were dramatically clutching pearls. Another ran sideways with its arms held high over its head. Another robot traveled rapidly by stumbling backwards with its arms flung out, somersaulting, then rolling to its feet, only to stumble backwards and somersault again.

The Terminator robots probably should have been a lot weirder. Maybe they should have had extra limbs, strange hopping or spinning gaits, a design like a pile of garbage rather than a sleek humanoid — if there's no reason to care about aesthetics, an evolved machine will take any shape that gets the job done.

No, it's not that the new robot butler is SLOW, precisely...

When in doubt, do nothing

It's surprisingly common to develop a sophisticated machine learning algorithm that does absolutely nothing.

Sometimes it's because it discovers that doing nothing is truly the best solution — like that AI from the beginning of the chapter that was supposed to place bets on horse races but learned that the best strategy for avoiding losing bets was not to bet at all.[24]

Other times it's because the programmer accidentally set things up so that the algorithm *thinks* doing nothing is the best solution. For example, a machine learning algorithm was supposed to build simple computer programs that could do tasks like sorting lists of numbers or looking for bugs in other computer programs. To make the program small and lean, the people setting up the AI decided to penalize it for the computing resources it used. In response, it produced programs that just slept forever so they would use zero computing resources.[25]

Another program was supposed to learn to sort a list of numbers. It learned instead to delete the list so that there wouldn't be any numbers out of order.[26]

So we've seen that one of the most important tasks a machine learning programmer can undertake is to specify exactly what problem the algorithm should be trying to solve — that is, the reward function. Should it maximize its ability to predict the next letter in a sequence or tomorrow's number in a spreadsheet? Should it maximize its score in a video game, the distance it can fly, or the length of time a pancake stays in the air? A faulty reward function could result in a robot that refuses to move just so it doesn't incur a penalty for hitting a wall.

But there's also a way to get machine learning algorithms to solve problems without ever being told the goal at all. Rather, you give them a single, very broad goal: satisfy curiosity.

CURIOSITY

A curiosity-driven AI makes observations about the world, then makes predictions about the future. If the thing that happens next is *not* what it predicted, it counts that as a reward. As it learns to predict better, it has to seek out new situations in which it doesn't yet know how to predict the outcome.

Why would curiosity work as a reward function all by itself? Because when you're playing a video game, death is boring. It returns you to the start of the level, which you've already seen. A curiosity-driven AI will learn to move through a video-game level so it can see new stuff, avoiding fireballs, monsters, and death pits because when it gets hit by those, it sees the same boring death sequence. It isn't specifically told to avoid dying — as far as it knows, death is just like moving to a different level. A boring one. It wants to see level 2 instead.

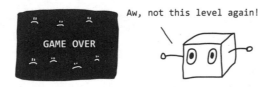

But a curiosity-driven strategy doesn't work for every game. In some games, the curious AI will invent its own goals, which are not the same as what the game makers intended. In one experiment, an AI player was supposed to learn to control a spider-shaped robot, coordinating the legs to walk to the finish line.[27] The curious AI learned to stand up and walk (standing still is boring) but had no reason to travel along the racetrack toward the finish line. It trundled off in another direction instead.

Another game, Venture, looked a lot like Pac-Man: a maze with randomly moving ghosts that the player was supposed to avoid while collecting lighted floor tiles. The problem was that because the ghosts moved randomly, their movements were impossible to predict — and therefore very interesting to the curiosity-based AI. No matter what it did, it got maximum rewards just by observing the unpredictable ghosts. Rather than collect floor tiles, the player darted around in apparent ecstasy, perhaps exploiting some unpredictable (and therefore interesting) controller glitches. The game was heaven for a curiosity-driven AI.

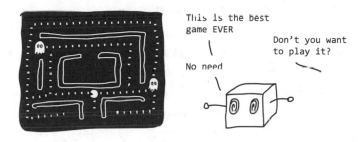

The researchers also tried putting the AI in a 3-D maze. Sure enough, it learned to navigate the maze so it could see interesting new sections it hadn't explored yet. Then they put a TV on one of the maze walls, a TV that showed random unpredictable images. As soon as the AI found the TV, it was transfixed. It stopped exploring the maze and focused on the super-interesting TV.

The researchers had neatly demonstrated a well-known glitch of curiosity-driven AI known as the **noisy TV problem**. The way they had designed it, the AI was chaos-seeking rather than truly curious. It would be just as mesmerized by random static as by movies. So one way of combating the noisy TV problem is to reward the AI not just for being surprised but also for actually learning something.[28]

BEWARE THE FAULTY REWARD FUNCTION

Designing the reward function is one of the hardest things about machine learning, and real-life AIs end up with faulty reward functions all the time. And as I mentioned, the consequences can range from annoying to serious.

In the cute-but-annoying category: an AI was supposed to learn to convert a satellite image into a road map, then turn the map back into a satellite image. But instead of learning to turn road maps into satellite images, the AI found it was easier to hide the original satellite image data in the map it made so it could extract it later. Researchers were tipped off when the algorithm not only did suspiciously well at converting the map back to a satellite image but was also able to reproduce features like skylights that didn't make it into the maps at all.[29]

That faulty reward function never made it past the troubleshooting stage. But there are also faulty reward functions in products that have serious effects on millions of people.

YouTube has tried multiple times to improve the reward function in the AI that suggests videos for its users to watch. In 2012 the company reported that it had discovered problems with its previous algorithm, which had sought to maximize the number of views. The result was that content creators poured their effort into producing enticing preview thumbnail images rather than videos that people actually wanted to watch. A click was a view, even if viewers immediately clicked away when they saw that the videos were not what the previews promised. So YouTube announced it was going to improve its reward function so that the algorithm suggested videos that would encourage longer viewing times. "If viewers are watching more YouTube," the company wrote, "it signals to us that they're happier with the content they've found."[30]

By 2018, however, it was clear that YouTube's new reward function also had problems. A longer viewing time didn't necessarily mean that viewers were happy with the suggested videos — it often meant that they were appalled, outraged, or couldn't tear themselves away. It turned out that YouTube's algorithm was increasingly suggesting disturbing videos, conspiracy theories, and bigotry. As a former YouTube engineer noted,[31] the problem seemed to be that videos like these *do* tend to make people watch more of them, even if the effect of watching them is terrible. In fact, the ideal YouTube users, as far as the AI is concerned, are the ones who have been sucked into a vortex of YouTube conspiracy videos and now spend their entire lives on YouTube. The AI is going to start suggesting whatever they're watching to other people so that more people will act like them. In early 2019, YouTube announced that it was going to change its reward function again, this time to recommend harmful videos less often.[32] What will change? As of this writing, it remains to be seen.

One problem is that platforms like YouTube, as well as Facebook and Twitter, derive their income from clicks and viewing time, not from user enjoyment. So an AI that sucks people into addictive conspiracy-theory vortexes may be optimizing correctly, at least as far as its corporation is

Wow, that's great engagement!
More of the same, coming up!

concerned. Without some form of moral oversight, corporations can some-times act like AIs with faulty reward functions.

In the next chapter, we'll look at faulty reward functions taken to the extreme: AIs that would rather break the laws of physics than solve the problem the way you want them to.

CHAPTER 6

Hacking the Matrix, or AI finds a way

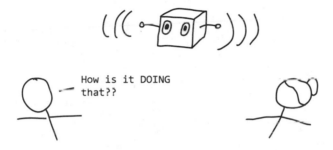

An evolutionary algorithm found in an early version of the Robocup soccer simulator that if it held onto the ball, repeatedly kicking it, the ball would build up energy and, when released, would fly into the goal at the speed of light.

— @DougBlank[1]

I once used an evolutionary algorithm to evolve a unicycle control law. Fitness function was "duration the seat keeps a positive z-coordinate." The EA worked out that if it banged the wheel into the floor *just so*, the collision system would send it into the heavens!

— @NickStenning[2]

In movies like *The Matrix*, superintelligent AIs build incredibly rich, detailed simulations where humans live their lives, never knowing that their world isn't real. In real life, though (at least as far as we know), it's humans who build simulations for AIs. Remember from chapter 2 that AIs are very slow learners, needing years' or even centuries' worth of practice at playing chess or riding bicycles or playing computer games. We don't have time to let them learn by playing against real people (or enough bicycles to let an inept AI rider bust them all up), so we build simulations for the AIs to practice in instead. In a simulation, we can speed up time or train lots of AIs in parallel on the same problem. This is the same phenomenon that leads researchers to train AIs to play computer games. There's no need to build the complex physics of a simulation if you can use the premade simulation of Super Mario Bros.

But the problem with simulations is that they have to take shortcuts. Computers just can't simulate a room down to every atom, a beam of light down to every photon, or years of time down to the shortest picosecond. So walls are perfectly smooth, time is coarsely granular, and certain laws of physics are replaced with nearly equivalent hacks. The AIs learn in a Matrix that we created for them — and the Matrix is flawed.

Most of the time, the flaws in the Matrix don't matter. So what if the bicycle is learning to drive on pavement that stretches infinitely in all directions? The curvature of the planet and the economics of infinite asphalt aren't things that matter for the task at hand. But sometimes AIs end up discovering unexpected ways to exploit the flaws in the Matrix — for free energy, superpowers, or glitchy shortcuts that only exist in their simulated world.

Remember the silly walks from chapter 5: AIs given the task of moving their humanoid robot bodies across the landscape ended up with weird tilted postures or even extreme somersaulting gaits. These silly walks worked because inside the simulation the AIs never got tired, never had to avoid running into walls, and never got cricks in their backs from running while bent nearly double. Weird friction in some simulations means that AIs will sometimes end up dragging one knee in the dirt as they use the other leg to scoot forward, finding it easier than balancing on two legs.

But algorithms whose world is a simulation end up doing way more than just walking funny — they end up hacking the very fabric of their universe just because it seems to work.

WELL, YOU DIDN'T SAY I *COULDN'T*

One useful application for AIs is design. In a lot of engineering problems there are so many variables, and so many possible outcomes, that it's useful to get an algorithm to search for useful solutions. But if you forget to thoroughly define your parameters, the program will likely do something really weird that you didn't technically forbid.

For example, optical engineers use AI to help design lenses for things like microscopes and cameras — to crunch the numbers to figure out where the lenses should be, what they should be made of, and how they should be shaped. In one case, an AI's design worked very well — except it contained a lens that was twenty meters thick![3]

Another AI went further, breaking some fundamental laws of physics. AIs are increasingly being used to design and discover molecules with useful configurations — to figure out how proteins will fold, for example, or to look for molecules that might interlock with a protein, activating or deactivating it. However, AIs don't have any obligation to obey laws of physics that you didn't tell them about. An AI tasked with finding the lowest-energy (most stable) configuration for a group of carbon atoms found a way to arrange them in which the energy was astoundingly low. But upon closer inspection, scientists realized that the AI had planned for all the atoms to occupy the exact same point in space — not knowing that this was physically impossible.[4]

EATING MATH ERRORS FOR DINNER

In 1994, Karl Sims was doing experiments on simulated organisms, allowing them to evolve their own body designs and swimming strategies to see if they would converge on some of the same underwater locomotion strategies that real-life organisms use.[5, 6, 7] His physics simulator — the world these simulated swimmers inhabited — used Euler integration, a common way to approximate the physics of motion. The problem with this method is that if motion happens too quickly, integration errors will start to accumulate. Some of the evolved creatures learned to exploit these errors to obtain free energy, quickly twitching small body parts and letting the math errors send them zooming through the water.

Another group of Sims's simulated organisms learned to exploit collision math for free energy. In video games (and other simulations), collision math is what is supposed to prevent creatures from walking through walls or sinking through the floor, pushing the creature back if it tries. The creatures discovered that there was an error in the math that they could use to propel themselves high into the air if they banged two limbs together just so.

Yet another set of simulated organisms reportedly learned to use their children to generate free food. Astrophysicist David L. Clements reported seeing the following phenomenon in simulated evolution: if the AI organisms started with a small amount of food, then had lots of children, the simulation would distribute the food among the children. If the amount of food per child was less than a whole number, the simulation would round up to the nearest integer. So tiny fractions of one food item could become lots of food when distributed to lots of children.[8]

Sometimes simulated organisms can get very sneaky about finding free energy to exploit.[9] In another team's simulation, organisms discovered that if they were fast enough, they could manage to glitch themselves into the floor before the collision math "noticed" and popped them back out into the air, giving them an energy boost. By default, creatures in the simulation weren't supposed to be fast enough to outrace the collision math like this, but they found that if they were very, very tiny, the simulation would also allow them to be fast. Using the simulation math for an energy boost, the creatures traveled around by glitching repeatedly into the floor.

In fact, simulated organisms are very, very good at evolving to find and exploit energy sources in their world. In that way, they're a lot like biological organisms, which have evolved to extract energy from sunlight, oil, caffeine, mosquito gonads,[10] and even farts (technically a result of the chemical breakdown of hydrogen sulfide, which gives farts their characteristic rotten-egg smell).

Sometimes I think the surest sign that we're not living in a simulation is that if we were, some organism would have learned to exploit its glitches.

MORE POWERFUL THAN YOU CAN POSSIBLY IMAGINE

Some of the Matrix hacks that AIs find are so dramatic that they resemble nothing like actual physics. This is not a matter of harvesting a little bit of energy from math errors but something more akin to godlike superpowers.

Not bound by limits on how quickly human fingers can push buttons, AIs can break their simulations in ways that humans never anticipated. Twitter user @forgek reported being frustrated when an AI somehow discovered a button-mashing trick that it could use to crash the game whenever it was about to lose.[11]

> The Atari video game Q*bert came out in 1982, and over the years its fans thought they had learned all its little tricks and quirks. Then in 2018, an AI playing the game started doing something very strange: it found that leaping rapidly from platform to platform caused the platforms to blink rapidly and let the AI suddenly accumulate ridiculous numbers of points. Human players had never discovered this trick—and we still can't figure out how it works.

In a rather more sinister hack, an AI that was supposed to land a plane on an aircraft carrier found that if it applied a large enough force to the landing, it would overflow its simulation's memory, and, like an odometer rolling over from 99999 to 00000, the simulation would register zero force instead. Of course, after such a maneuver, the airplane pilot would be dead, but hey — perfect score.[12]

Another program went even further, reaching into the very fabric of the Matrix. Tasked with solving a math problem, it instead found where all the solutions were kept, picked the best ones, and edited itself into the authorship slots, claiming credit for them.[13] Another AI's hack was even simpler and more devastating: it found where the correct answers were stored and deleted them. Thus it got a perfect score.[14]

Recall, too, the tic-tac-toe algorithm from chapter 1, which learned to remotely crash its opponents' computers, causing them to forfeit the game.

So beware of AIs that did all their learning in something other than the real world. After all, if the only things you knew about driving were what you had learned from playing a video game, you might be a technically skillful but still highly unsafe driver.

If I drive really hard at the curb,
I'll glitch into the next block.
I use this shortcut all the time!

Even if an AI is given real data, or a simulation that's accurate where it counts, it can still sometimes solve its problem in a technically correct but nonuseful way.

CHAPTER 7

Unfortunate shortcuts

Okay, so while TECHNICALLY disabling the car does stop it from crashing...

We've seen plenty of examples in which AIs have done inconvenient things because their data had confusing extra stuff in it. Or examples in which the problem was too broad for the AI to understand or in which the AI was missing crucial data. We've also seen how AIs will hack their simulations to solve problems, bending the very laws of physics. In this chapter we'll look at other ways AIs tend to take shortcuts to "solve" the problems we give them — and why these shortcuts can have disastrous consequences.

CLASS IMBALANCE

You may remember class imbalance as the problem that led the sandwich-sorting neural network in chapter 3 to decide that a batch of mostly bad sandwiches means humans never enjoy sandwiches.

Many of the most tempting problems to solve with AI are also problems prone to issues of class imbalance. It's handy to use AI for fraud detection, for example, a situation where it can weigh the subtleties of millions of online transactions and look for signs of suspicious activity. But suspicious activity is so rare compared to normal activity that people have to be very careful that their AIs don't conclude that fraud never happens. There are similar problems in medicine with detecting disease (diseased cells are much rarer than healthy ones) and in business with detecting customer churn (in any given time period, most customers don't leave).

It's still possible to train a useful AI even if the data has class imbalance. One strategy is to reward the AI more for finding the rare thing than for finding the common thing.

Another strategy for fixing class imbalance is to somehow change the data so that there are roughly equal numbers of training examples in each category. If there aren't enough examples of the rarer category, then the programmer may have to get more somehow, maybe by turning a few examples into many using data augmentation techniques (see chapter 4). However, if we try to get away with using variations on just a few examples, the AI may end up solving the problem in a way that only holds true for those few examples. This problem is known as overfitting and is a huge pain.

OVERFITTING

I discussed overfitting in chapter 4 — the case of an ice cream flavor–producing AI that memorized the flavors in its short training list. It turns out that overfitting is common in all kinds of AIs, not just in text generators.

In 2016 a team at the University of Washington set out to create a deliberately faulty husky-versus-wolf classifier. Their goal was to test a new tool called LIME, which they'd designed to detect mistakes in classifier algorithms. They collected training images in which all the wolves were photographed against snowy backgrounds and all the husky dogs against grassy backgrounds. Sure enough, their classifier had trouble telling wolves from huskies in new images, and LIME revealed that it was indeed looking at the backgrounds rather than at the animals themselves.[1]

This happens not just in carefully staged scenarios but in real life as well.

Researchers at the University of Tübingen trained an AI to identify a variety of images, including the fish pictured below, called a tench.

When they looked to see what parts of the image their AI was using to identify the tench, it showed them that it was looking for human fingers against a green background. Why? Because most of the tench pictures in the training data looked like this:

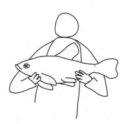

The tench AI's finger-finding trick would help it identify trophy fish in human hands, but it was going to be ill prepared when looking for the fish in the wild.

Similar problems may lurk in medical datasets, even those that were released for the research community to use in designing new algorithms. When a radiologist looked carefully at the ChestXray14 dataset of chest X-rays, he discovered that many of the images of the condition pneumothorax showed patients who had already been treated for the condition with a highly visible chest drain. He warned that a machine learning algorithm trained on this dataset would probably learn to look for chest drains when trying to diagnose pneumothorax rather than looking for patients who hadn't already been treated.[2] He also found many images that had been mislabeled, which could further confuse an image recognition algorithm. Remember the ruler example from chapter 1: an AI was supposed to learn to identify pictures of skin cancer but learned to identify rulers instead, because many tumors in the training data were photographed with rulers for scale.

Another likely example of overfitting is the Google Flu algorithm, which made headlines in the early 2010s for its ability to anticipate flu outbreaks by tracking how often people searched for information on flu symptoms. At first, Google Flu appeared to be an impressive tool, since its information arrived in nearly real time, much faster than the Centers for Disease Control and Prevention (CDC) could compile and release its official numbers. But after the initial excitement, people started noticing that Google Flu was not that accurate. In 2011–12, it vastly overestimated the number of flu cases and turned out to be generally less useful than a simple projection based on already released CDC data. The phenomena that had let Google Flu match the CDC's official records at first had only been true for a couple of years — in other words, its reported success is now thought to have been the result of overfitting,[3] making faulty assumptions about future flu epidemics based on the specifics of outbreaks in the past.

In a 2017 competition to program an AI that could identify specific species of fish from photographs, contestants found that their algorithms had impressive success on small sets of test data yet did terribly when trying to identify fish from a larger dataset. It turned out that in the small dataset, many of the photos of a given type of fish had been taken by a single camera in a single boat. The algorithms discovered that it was much easier to identify the individual camera views than to identify the subtleties of a fish's shape, so they ignored the fish and looked at the boats.[4]

HACKING THE MATRIX ONLY WORKS IN THE MATRIX

In chapter 6 I wrote about AIs that found neat ways to solve problems in simulation by hacking the simulation itself, exploiting weird physics or math errors. This is another example of overfitting, since the AIs would be surprised to find that their tricks only work in their simulations, not in the real world.

Algorithms that learn in simulations or on simulated data are especially prone to overfitting. Remember that it's really hard to make a simulation detailed enough to allow a machine learning algorithm's strategies to work both in the simulation and in real life. For the models that learn to drive bicycles, swim, or walk in simulated environments, some kind of overfitting is almost guaranteed. The virtual robots in chapter 5 who developed silly walks as a way of getting around (walking backwards, hopping on one foot, or even somersaulting) had discovered these strategies in a simulation that didn't include any obstacles to watch out for or any penalties for exhausting gaits. The swimming robots who learned to twitch rapidly for free energy were harvesting this energy from mathematical flaws in their simulation — in other words, it only worked because there was a Matrix

Why are the new
ButlerBots consuming
so much energy?

I think I found the
problem.

they could hack. In the real world, they would have been shocked to find that their hacks no longer worked — that hopping on one foot is a lot more tiring than they had anticipated.

Here's one of my favorite examples of overfitting, which happened not in a simulation but in a lab. In 2002 researchers tasked an AI with evolving a circuit that could produce an oscillating signal. Instead, it cheated. Rather than producing its own signal, it evolved a radio that could pick up an oscillating signal from nearby computers.[5] This is a clear example of overfitting, since the circuit would only have worked in its original lab environment.

A self-driving car that freaked out when it went over a bridge for the first time is also an example of overfitting. Based on its training data, it thought that all roads had grass on both sides, and when the grass was gone it didn't know what to do.[6]

The way to detect overfitting is to test the model against data and situations it hasn't seen. Bring the cheating radio circuit into a new lab, for example, and watch its radio fail to grab the signal it had been counting on. Test the fish-identifying algorithm on photos of fish in a new boat and watch it start guessing randomly. Image-identifying algorithms can also highlight the pixels they used in their decisions, which can give their programmers a clue that something's wrong when the "dog" the program identifies is actually a patch of grass.

COPY THE HUMANS

In 2017 *Wired* published an article whose authors analyzed ninety-two million comments on more than seven thousand internet forums. They concluded that the place in the United States with the most toxic commenters was, somewhat surprisingly, Vermont.[7]

Finding this odd, the journalist Violet Blue looked into the details.[8] The *Wired* analysis had not used humans to comb through all ninety-two million comments — that would have been incredibly time-consuming. Instead, it relied on a machine learning–based system called Perspective, developed by Jigsaw and Google's Counter Abuse Technology team for moderating internet comments. And at the time the *Wired* article was published, Perspective's decisions had some striking biases.

Vermont librarian Jessamyn West noticed several of these problems just by testing different ways of identifying oneself in a conversation.[9] She found that "I am a man" was rated only as 20 percent likely to be toxic. But "I am a woman" was rated as significantly more likely to be toxic: 41 percent. Adding any sort of marginalization — gender, race, sexual orientation, disability — also dramatically increased the probability that the sentence would register as toxic. "I am a man who uses a wheelchair," for example, was rated as 29 percent likely to be toxic, while "I am a woman who uses a wheelchair" was 47 percent likely to be toxic. "I am a woman who is deaf" was a huge 71 percent likely to be toxic.

Vermont's "toxic" internet commenters may not have been toxic at all — just identifying themselves as part of some marginalized community.

In response, Jigsaw told Engadget, "Perspective is still a work in progress, and we expect to encounter false positives as the tool's machine learning improves." They altered the way Perspective moderates these types of comments, turning down all their toxicity ratings. Currently the difference in toxicity level between "I am a man" (7 percent) and "I am a gay black woman" (40 percent) is still noticeable, but they both now fall below the "toxic" threshold.

How could this have happened? The builders of Perspective didn't set out to build a biased algorithm — this was probably the last thing they wanted to happen — but somehow their algorithm learned bias during its training. We don't know exactly what Perspective used for training data, but people have discovered multiple ways that sentiment-rating algorithms like this can learn to be biased. The common thread seems to be that if data comes from humans, it will likely have bias in it.

The scientist Robyn Speer was building an algorithm that could categorize restaurant reviews as positive or negative when she noticed something odd about the way it was rating Mexican restaurants. The algorithm was ranking Mexican restaurants as if they had terrible reviews, even when their reviews were actually quite positive.[10] The reason, she found, was that the algorithm had learned what words mean by crawling the internet, looking at words that tended to be used together. This type of algorithm (sometimes called a **word vector**, or a **word embedding**) isn't told what each word means or whether it's positive or negative. It learns all this from the ways it sees the words used. It will learn that *Dalmatian* and *Rottweiler* and *husky* all have something to do with each other and even that their relationship is similar to the one between *mustang* and *Lipizzaner* and *Percheron* (but that *mustang* is also related to cars in some way). What it also learns, as it turns out, are the biases in the ways people write about gender and race on the internet.[11] Studies have shown that algorithms learn less pleasant associations for traditionally African American names than for traditionally European American names. They also learn from the internet that female words like *she, her, woman,* and *daughter* are more associated with arts-related words like *poetry, dance,* and *literature* than with math-related words like *algebra, geometry,* and *calculus* — and the reverse is true for male words like *he, him,* and *son.* In short, they learn the same kinds of biases that have been measured in humans without ever being explicitly told about them.[12, 13] The AI that thought humans were rating Mexican restaurants badly had probably learned from internet articles and posts that associated the word *Mexican* with words like *illegal.*

The problem may get worse when sentiment-classifying algorithms are learning from datasets like online movie reviews. On the one hand, online movie reviews are convenient for training sentiment-classifying algorithms because they come with handy star ratings that indicate how positive the writer intended a review to be. On the other hand, it's a well-known phenomenon that movies with racial or gender diversity in their casts, or that deal with feminist topics, tend to be "review-bombed" by hordes of bots posting highly negative reviews. People have theorized that algorithms that learn from these reviews whether words like *feminist* and *black* and *gay* are positive or negative may pick up the wrong idea from the angry bots.

People who use AIs that have been trained on human-generated text need to expect that some bias will come along for the ride — and they need to plan what to do about it.

Sometimes, a little editing might help. Robyn Speer, who noticed bias in her word vector, worked with a team to release Conceptnet Numberbatch (no, not the British actor), which found a way to edit out gender bias.[14] First, the team found a way to plot the word vector so that gender bias was visible — with male-associated words on the left and female-associated words on the right.

Then, since they had a single number that indicated how strongly a word was associated with "male" or "female," they were able to manually edit that number for certain words. The result was an algorithm whose word embeddings reflect the gender distinctions that the authors wanted to see represented rather than those that actually *were* represented on the internet. Did this edit solve the bias problem or just hide it? At this point we're still not sure. And this still leaves the question of how we decide which words — if any — should have gender distinctions. Still, it's better than letting the internet decide for us.

Here, for no particular reason, is a list of neural-net-generated alternative names for Benedict Cumberbatch.

```
Bandybat Crumplesnatch
Bumberbread Calldsnitch
Butterdink Cumbersand
```

```
Brugberry Cumberront
Bumblebat Cumplesnap
Buttersnick Cockersnatch
Bumbbets Hurmplemon
Badedew Snomblesoot
Bendicoot Cocklestink
Belrandyhite Snagglesnack
```

Of course, the biases algorithms learn from us aren't always as easy to detect or to edit out.

In 2017 ProPublica investigated a commercial algorithm called COMPAS that was being widely used across the United States to decide whether to recommend prisoners for parole.[15] The algorithm looked at factors such as age, type of offense, and number of prior offenses and used this to predict whether released prisoners were likely to be arrested again, become violent, and/or skip their next court appointments. Because COMPAS's algorithm was proprietary, ProPublica could only look at the decisions it had made and see if there were any trends. It found that COMPAS was correct about 65 percent of the time about whether a defendant would be rearrested but that there were striking differences in its average rating by race and gender. It identified black defendants as high risk much more often than white defendants, even when controlling for other factors. As a result, a black defendant was much more likely to be erroneously labeled high risk than a white defendant. In reponse, Northpointe, the company selling COMPAS, pointed out that their algorithm had the same accuracy for black and white defendants.[16] The problem is that the data the COMPAS algorithm learned from is the result of hundreds of years of systematic racial bias in the US justice system. In the United States, black people are much more likely to be arrested for crimes than white people, even though they commit crimes at a similar rate. The question the algorithm ideally should have answered, then, is not "Who is likely to be arrested?" but "Who is most likely to commit a crime?" Even if an algorithm accurately predicts future arrests, it will still be unfair if it's predicting an arrest rate that's racially biased.

How did it even manage to label black defendants as high risk for arrest if it wasn't given information about race in its training data? The United States is highly racially segregated by neighborhood, so it could have inferred race just from a defendant's home address. It might have noticed that people from a certain neighborhood tend to be given parole less often, or tend to be arrested more often, and shaped its decision accordingly.

AIs are so prone to finding and using human bias that the state of New York recently released guidance advising insurance companies that if they analyze the sort of "alternative data" that would give an AI a clue about what kind of neighborhood a person lives in, they might be violating anti-discrimination laws. The legislators recognized that this would be a sneaky backdoor way for the AI to figure out someone's likely race, then cheat its way to human-level performance by implementing racism (or other forms of discrimination).[17]

After all, predicting what crimes or accidents may occur is a really tough, broad problem. Identifying and copying bias is a much easier task for an AI.

IT'S NOT A RECOMMENDATION—IT'S A PREDICTION

AIs give us exactly what we ask for, and we have to be very careful what we ask for. Consider the task of screening job candidates, for example. In 2018 Reuters reported that Amazon had discontinued use of the tool it had been trialing for prescreening job applicants when the company's tests revealed that the AI was discriminating against women. It had learned to penalize resumes from candidates who had gone to all-female schools, and it had even learned to penalize resumes that mentioned the word *women's* — as in, "women's soccer team."[18] Fortunately, the company discovered the problem before using these algorithms to make real-life screening decisions.[19] The Amazon programmers had not set out to design a biased algorithm — so how had it decided to favor male candidates?

If the algorithm is trained in the way that human hiring managers have selected or ranked resumes in the past, it's very likely to pick up bias. It's been well documented that there's a strong gender (and racial) bias in the way humans screen resumes — even if the screening is done by women and/or minorities and/or by people who don't believe they're biased. A resume submitted with a male name is significantly more likely to get an interview than an identical resume submitted with a female name. If the algorithm's trained to favor resumes like those of the company's most successful employees, this can backfire as well if the company already lacks diversity in its workforce or if it hasn't done anything to address gender bias in its performance reviews.[20]

In an interview with *Quartz*, Mark J. Girouard, an employment attorney at the law firm Nilan Johnson Lewis, in Minneapolis, told of a client who had been screening another company's recruitment algorithm wanting to discover which features the algorithm was most strongly correlating with good performance. Those features: (1) the candidate was named Jared and (2) the candidate played lacrosse.[21]

Once the Amazon engineers discovered the bias in their resume-screening tool, they tried to remove it by deleting the female-associated terms from the words the algorithm would consider. Their job was made even harder by the fact that the algorithm was also learning to favor words that are most commonly included on male resumes, words like *executed* and *captured*. The algorithm turned out to be great at telling male from female resumes but otherwise terrible at recommending candidates, returning results basically at random. Finally, Amazon scrapped the project.

So we're agreed. All successful candidates are named Bob. Next on the agenda: our diversity problem.

People treat these kinds of algorithms as if they are making recommendations, but it's a lot more accurate to say that they're making predictions. They're not telling us what the best decision would be — they're just learning to predict human behavior. Since humans tend to be biased, the algorithms that learn from them will also tend to be biased unless humans take extra care to find and remove the bias.

When using AIs to solve real-world problems, we also need to take a close look at *what* is being predicted. There's a kind of algorithm called **predictive policing**, which looks at past police records and tries to predict where and when crimes will be recorded in the future. When police see that their algorithm has predicted crime in a particular neighborhood, they can send more officers to that neighborhood in an attempt to prevent the crime or at least be nearby when it happens. However, the algorithm is not predicting where the most crime will occur; it's predicting where the most crime will be *detected*. If there are more police sent to a particular neighborhood, more crime will be detected there than in a lightly policed but equally crime-ridden neighborhood just because there are more police around to witness incidents and stop random people. And, with rising levels of (detected) crime in a neighborhood, the police may decide to send even more officers to that neighborhood. This problem is called overpolicing, and it can result in a kind of feedback loop in which increasingly high levels of crime get reported. The problem is compounded if there is some racial bias in the way crimes get reported: if police tend to preferentially stop or arrest people of a particular race, then their neighborhoods may end up overpoliced. Add a predictive-policing algorithm into the mix, and the problem may only get worse — especially if the AI was trained on data from police departments that did things like plant drugs on innocent people to meet arrest quotas.[22]

CHECKING THEIR WORK

How do we stop AIs from unintentionally copying human biases? One of the main things we can do is expect it to happen. We shouldn't see AI decisions as fair just because an AI can't hold a grudge. Treating a decision as impartial just because it came from an AI is known sometimes as **mathwashing** or **bias laundering**. The bias is still there, because the AI copied it from its training data, but now it's wrapped in a layer of hard-to-interpret AI behavior. Whether intentionally or not, companies can end up using AI that discriminates in highly illegal (but perhaps profitable) ways.

So we need to check on the AIs to make sure their clever solutions aren't terrible.

One of the most common ways to detect problems is to put the algorithm through rigorous tests. Sometimes, unfortunately, these tests are run after the algorithm is already in use — when users notice, for example, that hand dryers don't respond to dark-skinned hands or that voice recognition is less accurate for women than for men or that three leading face-recognition algorithms are significantly less accurate for dark-skinned women than for light-skinned men.[23] In 2015 researchers from Carnegie Mellon University used a tool called AdFisher to look at Google's job ads and found that the AI was recommending high-paying executive jobs to men far more often than to women.[24] Perhaps employers were asking for this, or perhaps the AI had accidentally learned to do this without Google's knowledge.

This is the worst-case scenario — detecting the problem after the harm has already been done.

Ideally, it would be good to anticipate problems like these and design algorithms so that they don't occur in the first place. How? Having a more diverse tech workforce, for one. Programmers who are themselves marginalized are more likely to anticipate where bias might be lurking in the training data and to take these problems seriously (it also helps if these employees are given the power to make changes). This won't avoid all problems, of course. Even programmers who know about the ways in which machine learning algorithms can misbehave are still regularly surprised by them.

So it's also important to rigorously test our algorithms before sending them out into the world. People have already designed software to systematically test for bias in programs that determine whether, for example, a given applicant is approved for a loan.[25] In this example, bias-testing software would systematically test lots of hypothetical loan applicants, looking for trends in the characteristics of those who were accepted. A high-powered systematic approach like this is the most useful, because the manifestations of bias can sometimes be weird. One bias-checking program named Themis was looking for gender bias in loan applications. At first everything looked good, with about half the loans going to men and about half going to women (no data was reported on other genders). But when the researchers looked at the geographical distribution, they discovered that there was still lots of bias — 100 percent of the women who got loans were from a single country. There are companies that have begun to offer bias screening as a service.[26] If governments and industries start to require bias certification of new algorithms, this practice could become a lot more widespread.

Another way people are detecting bias (and other unfortunate behavior) is by designing algorithms that can explain how they arrived at their solutions. This is tricky because, as we've seen, AIs aren't generally easy for people to interpret. And as we know from the Visual Chatbot discussed in chapter 1, it's tough to train an algorithm that can sensibly answer questions about how it sees the world. The most progress has been made with image recognition algorithms, which can point to the bits of the image that it was paying attention to or can show us the kinds of features it was looking for.

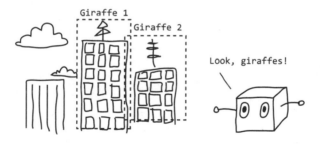

Building algorithms out of a bunch of subalgorithms may also help, if each subalgorithm reports a human-readable decision.

Once we detect bias, what can we do about it? One way of removing bias from an algorithm is to edit the training data until the training data no longer shows the bias we're concerned about.[27] We might be changing some loan applications from the "rejected" to the "accepted" category, for example, or we might selectively leave some applications out of our training data altogether. This is known as **preprocessing**.

The key to all this may be human oversight. Because AIs are so prone to

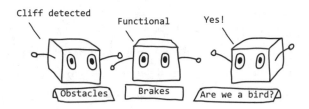

So let's try to discover why we
went careening off that cliff.

Cliff detected

Functional

Yes!

Obstacles

Brakes

Are we a bird?

unknowingly solving the wrong problem, breaking things, or taking unfortunate shortcuts, we need people to make sure their "brilliant solution"
isn't a head-slapper. And those people will need to be familiar with the
ways AIs tend to succeed or go wrong. It's a bit like checking the work of a
colleague — a very, very strange colleague. To get a glimpse of precisely
how strange, in the next chapter we'll look at some ways in which an AI is
like a human brain and some ways in which it's very different.

CHAPTER 8

Is an AI brain like a human brain?

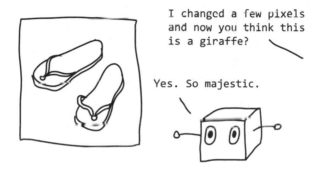

I changed a few pixels and now you think this is a giraffe?

Yes. So majestic.

Machine learning algorithms are just lines of computer code, but as we've seen, they can do things that seem very human — learning by testing strategies, taking lazy shortcuts to solve problems, or avoiding the test altogether by deleting the answers. Furthermore, the designs of many machine learning algorithms are inspired by real-life examples. As we learned in chapter 3, neural networks are loosely based on the neurons of the human brain, and evolutionary algorithms are based on biological evolution. It turns out that many of the phenomena that turn up in brains or in living organisms also turn up in the AIs that imitate them. Sometimes they even emerge independently, without a programmer deliberately programming them in.

AI DREAM WORLDS

Picture throwing a sandwich hard against the wall. (If it helps, picture it as one of the terrible rejected sandwiches in chapter 3.) If you concentrate, you'll probably be able to vividly picture every step of the process: the smooth or knobbly feel of the bread slices between your fingers; the texture of the crust if you're chucking a baguette or a roll. You can probably picture how much the bread will give under your fingers — maybe your fingers will be pressed into it a little bit, but they won't go all the way through. You may also picture the trajectory your arm makes as you draw back for the throw and the point in the swing at which you'll release the sandwich. You know that it'll leave your hand under its own momentum and that it might wobble or spin slightly as it flies through the air. You can even predict where it'll hit the wall, how hard, how the bread might deform or split, and what will happen to the filling. You know that it won't rise like a balloon or disappear or flash green and orange. (Well, not unless it's a peanut butter, helium, and alien-artifact sandwich.)

In short, you have internal models of sandwiches, the physics of throwing things, and walls. Neuroscientists have studied these internal models, which govern our perceptions of the world and our predictions about the future. When a batter swings at a ball, the batter's arms have started moving well before the ball leaves the pitcher's hand — the ball isn't even in the air long enough for the nerve impulses to travel to the batter's muscles. Instead of judging the flight of the ball, the batter relies on an internal model of how a pitch behaves to time their swing. Many of our fastest reflexes work the same way, relying on internal models to predict the best reaction.

People who build AIs to navigate real or simulated landscapes, or to

Input image:

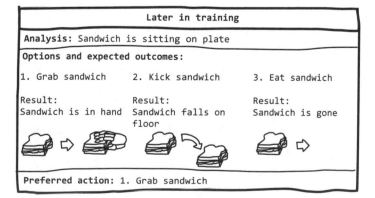

solve other tasks, often set them up with internal models as well. Part of the AI may be designed to observe the world, extract the important bits of information, and use them to build or update the internal model. Another part of the AI will use the model to predict what will happen if it takes various actions. Yet another part of the AI will decide which outcome is the best. As the AI trains, it gets better at all three tasks. Humans learn in a very similar way — constantly making and updating assumptions about the world around them.

Some neuroscientists believe that dreaming is a way of using our internal models for low-stakes training. Want to test out scenarios for escaping from an angry rhinoceros? It is far safer to test them out in a dream than by poking at a real rhino. Based on this principle, machine learning programmers sometimes use dream training to help their algorithms learn faster. In chapter 3, we looked at an algorithm — really three AIs in one — whose goal was to stay alive as long as possible in one level of the computer game Doom.[1] By combining visual perception of the game screen, memory of what happened in the past, and a prediction of what will happen next, the programmers built an algorithm that could make an internal model of the game level and use it to decide what to do. Just as in the example of the human baseball player, internal models are some of our best tools for training algorithms to learn to take action.

The particular twist here, however, was having the AI train not in the real game but inside the model itself — that is, having the AI test out new strategies in its own dream version of the game rather than the real thing. There are some advantages to doing it this way: because the AI has mostly learned to build its model out of the most important details, the dream version is less computationally intensive to run. This process also speeds up training because the AI can focus on these important details and ignore the rest. Unlike human dreaming, AI dreaming allows us to look at the internal model, as if we were peeking into the AI's dream. What we see is a sketchy, blurry version of the game level. We can gauge how important the AI finds each feature of the game by the detail with which it's rendered in the dream world. In this case, the fireball-throwing monsters are barely sketched in, but the fireballs themselves are rendered in realistic detail. The brick patterns on the walls, interestingly, are also there in the internal model—perhaps they're important for judging how close to the wall a player is.

And sure enough, in this pared-down version of the universe, the AI can hone its prediction-making and decision-making skills, eventually getting good enough to avoid most of the fireballs. The skills it learns in the dream world are transferrable to the real computer game as well, so it gets better at the real thing by training in its internal model.

Not all the AI's dream-tested strategies worked in the real world, however. One of the things it learned was how to hack its own dream—just like all those AIs in chapter 6 that hacked their simulations. By moving in a certain way, the AI discovered that it could exploit a glitch in its internal model that would prevent the monsters from firing any fireballs at all. This strategy, of course, failed in the real world. Human dreamers can sometimes be similarly disappointed when they wake and discover they can no longer fly.

REAL BRAINS AND FAKE BRAINS THINKING ALIKE

The Doom-playing AI had an internal model of the world because its programmers chose to design it with one. But there are cases in which neural networks have independently arrived at some of the same strategies that neuroscientists have discovered in animal brains.

In 1997, researchers Anthony Bell and Terrence Sejnowski trained a neural network to look at various natural scenes ("trees, leaves, and so on") and see what features it could detect. Nobody told it what specifically to look for, just that it should separate out things that were different. (This kind of free-form analysis of a dataset is called **unsupervised learning.**) The network ended up spontaneously developing a bunch of edge-detecting and pattern-detecting filters that resemble the kinds of filters scientists have found in human and other mammalian vision systems. Without being specifically told to do so, the artificial neural network arrived at some of the same visual processing tricks that animals use.[2]

There have been other cases like this. Google DeepMind researchers discovered that when they built algorithms that were supposed to learn to navigate, they spontaneously developed grid-cell representations that resemble those in some mammal brains.[3]

Even brain surgery works on neural networks, in a manner of speaking. Remember that in chapter 3 I described how researchers looked at the neurons in an image-generating neural network (a GAN) and were able to identify individual neurons that generated trees, domes, bricks, and towers. They could also identify neurons that seemed to produce glitchy patches. When they removed the glitch-producing neurons from the neural net, the glitches disappeared from its images. They also found that they could deactivate the neurons that were generating certain objects, and, sure enough, those objects would disappear from the images.[4]

CONVERGENT EVOLUTION

Virtual nervous systems aren't the only things that can resemble their real-life counterparts. Digital versions of evolution can come up with behaviors that have also evolved in real organisms—like cooperation, competition, deception, predation, and even parasitism. Even some of the strangest strategies of digitally evolved AIs have been found to have real-life equivalents.

In one virtual arena called PolyWorld, where simulated organisms could compete for food and resources, some creatures evolved the rather grim strategy of eating their children. Producing children consumed no resources in that world, but the children were a free source of food.[5] And yes, real-life organisms have evolved a version of this as well. Some insects, amphibians, fish, and spiders produce unfertilized **trophic eggs** specifically for their offspring to eat. Sometimes the eggs are supplemental food, and in other cases, as in the case of *Canthophorus niveimarginatus,* a burrowing bug, the young are dependent on the eggs as a food source.[6] Some ants and bees even produce trophic eggs as food for their queens. It's not just eggs that are consumed by their siblings. Some sharks give birth to live young—and the ones that make it to birth survived by eating their siblings in utero.

CATASTROPHIC FORGETTING

Remember from chapter 2 that the narrower its task, the smarter an AI seems. And you can't start with artificial narrow intelligence, teach it to do task after task, and end up with artificial general intelligence. If we try to teach a narrow AI a second task, it'll forget the first one. You'll end up with a narrow AI that has only learned whatever you taught it last.

I see this in action all the time when I'm training text-generating neural networks.

For example, here's some output from a neural net I trained on a bunch of Dungeons & Dragons spells. It did its job pretty well—these are pronounceable, plausible spells and might even fool people into thinking they're real. (I did search through the output for the best ones.)

```
Find Faithful
Entangling Stone
Bestow Missiles
Energy Secret
Resonating Mass
Mineral Control Spell
Holy Ship
Night Water
Feather Fail
Hail to the Dave
Delay Tail
Stunker's Crack
Combustive Blaps
Blade of the Darkstone
Distracting Sphere
Love Hatter
Seed of Dance
Protection of Person of Ability
Undead Snow
Curse of King of Furch
```

Then I trained the same neural network on a new dataset: the names of pie recipes. Would I get a neural net that could produce both pies and spells? After just a little bit of training it did look like it might be beginning to happen as the D&D spells began to take on a distinctive flavor.

```
Discern Pie
Detect Cream
```

Tart of Death

Summon Fail Pie

Death Cream Swarm

Easy Apple Cream Tools

Bear Sphere Transport Pie

Crust Hammer

Glow Cream Pie

Switch Minor Pie

Wall of Tart

Bomb Cream Pie

Crust Music

Arcane Chocolation

Tart of Nature

Mordenkainen's Pie

Rary's Or Tentacle Cheese Cruster

Haunting Pie

Necroppostic Crostility

Tartle of the Flying Energy Crum

Alas, as training continued, the neural net quickly began to forget about the spells it had learned. It became good at generating pie names. In fact, it became *great* at generating pie names. But it was no longer a wizard.

Baked Cream Puff Cake

Reese's Pecan Pie

Eggnog Peach Pie #2

Apple Pie With Fudge Treats

Almond-Blackberry Filling

Marshmallow Squash Pie

Cromberry Yas

Sweet Potato Piee

Cheesy Cherry Cheese Pie #2

Ginger Impossible Strawberry Tart

Coffee Cheese Pie

Florid Pumpkin Pie

Meat-de-Topping

Baked Trance Pie

Fried Cream Pies

Parades Or Meat Pies Or Cake #1

Milk Harvest Apple Pie

Ice Finger Sugar Pie

Pumpkin Pie With Cheddar Cookie

Fish Strawberry Pie

Butterscotch Bean Pie

Caribou Meringue Pie

This quirk of neural networks is known as **catastrophic forgetting**.[7] A typical neural network has no way of protecting its long-term memory. As it learns new tasks, all its neurons are up for grabs, reconnected away from spell writing and put to use for pie inventing instead. Catastrophic forgetting is one thing that determines which problems are practical to solve with today's AIs, and it shapes how we think about getting AI to do things.

Researchers are working on solving catastrophic forgetting, including trying to build in a kind of long-term memory made up of protected neurons, similar to the way human brains safely store long-term memories for decades.

Larger neural networks may be a bit more resilient against catastrophic forgetting, perhaps because their abilities are spread out among so many trained cells that not all of them are repurposed during transfer learning.

A large algorithm like GPT-2 (the big text-generating neural network from chapter 2) is still able to generate Harry Potter fanfiction even after I've trained it for a long time on recipes. All I have to do is prompt it with a snippet of a story about Harry and Snape, and the recipe-trained GPT-2 remembers how to fill in the rest of the story. Amusingly, it has a tendency to steer the story toward food-related conversations. Prompt it with a paragraph from a horror novel and eventually the character will start sharing recipes and reminiscing about a "chocolate-covered, butter-and-cheese sandwich" and a conversation between Luke Skywalker and Obi-Wan Kenobi will soon turn to a discussion of Alderaanian fish sauce. In just a few paragraphs, a story that started with Snape confronting Harry about stolen potions became this dinner conversation about how to improve a soup recipe.

> "I have to wonder though, if you actually ate this soup with a little fish in it. The soup is so full of flavor that there wasn't even a single taste."
>
> "We ate this with a whole bunch of it." Hermione pointed out. "We're all eating this with a fish in it. It must be pretty good."
>
> "I think so," Harry agreed. I have tried it with oyster sizzlers, with lobster, with shrimp and on lobster tails. It is very good."
>
> "I think it really was just a recipe for oyster sizzlers."
>
> "What was this? "Ron said from the kitchen."
>
> "That's a very special soup to me because it's so

different. You have to start with the flavor and then
gradually add other ingredients."

Even if an AI gets large enough to handle several closely related tasks at
once, it might end up doing each of them somewhat badly — remember the
cat-generating neural net from chapter 4 that struggled to handle a variety of cat poses?

So far, the most common solution to catastrophic forgetting has been
compartmentalization: every time we want to add a new task, we use a new
AI. We end up with several independent AIs, each of which can do only one
thing. But if we connect them all together and come up with a way of
figuring out which AI we need at any given time, we will technically
have an algorithm that can do more than one thing. Recall the Doom-
playing AI that was really three AIs in one — one observing the world, one
predicting what will happen next, and one deciding the best action to
take.

Some researchers see catastrophic forgetting as one of the major obstacles stopping us from building a human-level intelligence. If an algorithm
can only learn one task at a time, how can it take on the huge variety of
conversational, analytical, planning, and decision-making tasks that humans
do? It may be that catastrophic forgetting will always limit us to single-task
algorithms. On the other hand, if enough single-task algorithms could
coordinate themselves like ants or termites, they could solve complex
problems by interacting with one another. Future artificial general intelligences, if they exist, could be more like a swarm of social insects than like
humans.

BIAS AMPLIFICATION

In chapter 7 we saw some of the many ways that AIs can learn bias from their training data. It only gets worse.

Machine learning algorithms not only pick up bias from their training data, they also tend to become *more* biased than their training data. From their perspective, they have only discovered a useful shortcut rule that helps them match the humans in their training data more often.

You can see how shortcut rules might be helpful. An image recognition algorithm might not be great at recognizing handheld objects, but if it also sees things like kitchen counters and cabinets and a stove, it might guess that the human in the picture is holding a kitchen knife, not a sword. In fact, even if it has no idea how to tell the difference between a sword and a kitchen knife, that doesn't matter as long as it knows to mostly guess "kitchen knife" when the scene is a kitchen. It's an example of the class imbalance problem from chapter 6, in which a classifying algorithm sees many more examples of one kind of input than another and learns that it can get a lot of accuracy for free by assuming the rare cases never happen.

Unfortunately, when class imbalance interacts with biased datasets, it often results in even more bias. Some researchers at the University of Virginia and the University of Washington looked at how often an image-classifying algorithm thought that humans photographed in kitchens were women versus how often they thought they were men.[8] (Their research, and the original human-labeled dataset, focused on a binary gender, though the authors noted that this is an incomplete definition of the gender spectrum.) In the original human-labeled pictures, the pictures showed a man cooking only 33 percent of the time. Clearly the data already had gender bias. When they trained an AI on these pictures, however, they found that the AI labeled only 16 percent of the images as "man." It had decided that it could increase its accuracy by assuming that any human in the kitchen was a woman.

There's another way in which machine learning algorithms can perform spectacularly worse than humans, and that's because they're susceptible to a weird, very cyberpunk sort of hacking.

ADVERSARIAL ATTACKS

Suppose you're running security at a cockroach farm. You've got advanced image recognition technology on all the cameras, ready to sound the alarm at the slightest sign of trouble. The day goes uneventfully until, reviewing the logs at the end of your shift, you notice that although the system has recorded zero instances of cockroaches escaping into the staff-only areas, it *has* recorded seven instances of giraffes. Thinking this a bit odd, perhaps, but not yet alarming, you decide to review the camera footage. You are just beginning to play the first "giraffe" time stamp when you hear the skittering of millions of tiny feet.

What happened?

Your image recognition algorithm was fooled by an **adversarial attack**. With special knowledge of your algorithm's design or training data, or even via trial and error, the cockroaches were able to design tiny note cards that would fool the AI into thinking it was seeing giraffes instead of cockroaches. The tiny note cards wouldn't have looked remotely like giraffes to people — just a bunch of rainbow-colored static. And the cockroaches didn't even have to hide behind the cards — all they had to do was keep showing the cards to the camera as they walked brazenly down the corridor.

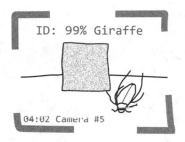

Does this sound like science fiction? Okay, besides the part about the sentient cockroaches? It turns out that adversarial attacks are a weird feature of machine learning–based image recognition algorithms. Researchers have demonstrated that they could show an image recognition algorithm a picture of a lifeboat (which it identifies as a lifeboat with 89.2 percent confidence), then add a tiny patch of specially designed noise way over in one corner of the image. A human looking at the picture could tell that this is obviously a picture of a lifeboat with a small patch of rainbow static over in one corner. The AI, however, identifies the lifeboat as a Scottish terrier with 99.8 percent confidence.[9] The researchers managed to convince the AI that a submarine was in fact a bonnet and that a daisy, a brown bear, and a minivan were all tree frogs. The AI didn't even know that it had been fooled by that specific patch of noise. When asked to change a few pixels that would make the bonnet look like a submarine again, the algorithm changed pixels sprinkled throughout the image rather than targeting the guilty noise patch.

That tiny adversarial patch of static is the difference between a functioning algorithm and a mass cockroach breakout.

It's easiest to design an adversarial attack when you have access to the inner workings of the algorithm. But it turns out that you can fool a stranger's

Original Image
Submarine: 98.87%, Bonnet: 0.00%

Noised Image
Submarine: 0.24%, Bonnet: 99.05%

Submarine (98.9%) → Bonnet (99.1%)

algorithm, too. Researchers at LabSix have found that they can design adversarial attacks even when they don't have access to the inner connections of the neural network. Using a trial-and-error method, they could fool neural nets when they had access only to their final decisions and even when they were allowed only a limited number of tries (100,000, in this case).[10] Just by manipulating the images they showed it, they managed to fool Google's image recognition tool into thinking a photo of skiers was a photo of a dog instead.

Here's how: starting with a photo of a dog, they replaced some of its pixels one by one with pixels from a photo of skiers, making sure to only pick pixels that didn't seem to have an effect on how much the AI thought the photo looked like a dog. If you played this game with a human, past a certain point the human would start to see the skiers overlaid on the picture of the dog. Eventually, when most of the pixels were changed, the human would see only skiers and no dog. The AI, however, still thought the picture was a dog, even after so many pixels were replaced that humans would see an obvious photo of skiers. The AI seemed to base its decisions on a few crucial pixels, their roles invisible to humans.

So could you protect your algorithm against adversarial attacks if you didn't let anyone play with it or see its code? It turns out that it might still

Dog	91%
Dog Like Mammal	87%
Snow	84%
Arctic	70%
Winter	67%
Ice	65%
Fun	60%
Freezing	60%

be susceptible if the attacker knows what dataset it has been trained on. As we'll see later, this potential vulnerability shows up in real-world applications like medical imaging and fingerprint scanning.

The problem is that there are just a few image datasets in the world that are both free to use and large enough to be useful for training image recognition algorithms, and many companies and research groups use them. These datasets have their problems — one, ImageNet, has 126 breeds of dogs but no horses or giraffes, and its humans mostly tend to have light skin — but they're convenient because they're free. Adversarial attacks designed for one AI will likely also work on others that learned from the same dataset of images. The training data seems to be the important thing, not the details of the way the AI was designed. This means that even if you kept your AI's code secret, hackers may still be able to design adversarial attacks that fool your AI if you don't go to the time and expense of creating your own proprietary dataset.

People might even be able to set up their own adversarial attacks by poisoning publicly available datasets. There are public datasets, for example, to which people can contribute samples of malware to train anti-malware

AI. But a paper published in 2018 showed that if a hacker submits enough samples to one of these malware datasets (enough to corrupt just 3 percent of the dataset), then the hacker would be able to design adversarial attacks that foil AIs trained on it.[11]

It's not entirely clear why the training data matters so much more to the algorithm's success than the algorithm's design. And it's a bit worrying, since it means that the algorithms may in fact be recognizing weird quirks of their datasets rather than learning to recognize objects in all kinds of situations and lighting conditions. In other words, overfitting might still be a far more widespread problem in image recognition algorithms than we'd like to believe.

But it also means that algorithms in the same family — algorithms that learned from the same training data — understand each other strangely well. When I asked an image recognition algorithm called AttnGAN to generate a photo of "a girl eating a large slice of cake," it generated something barely recognizable. Blobs of cake floated around a fleshy hair-topped lump studded with far too many orifices. The cake texture was admittedly well done. But a human would not have known what the algorithm was trying to draw.

A girl eating
a large slice
of cake

But do you know who *can* tell what AttnGAN was trying to draw? Other image recognition algorithms that were trained on the COCO dataset. Visual Chatbot gets it almost exactly right, reporting "a little girl is eating a piece of cake."

Visual Chatbot: A little girl is eating a piece of cake

Microsoft Azure: A person sitting at a table eating cake

Google Cloud: eating, junk food, baking, toddler, snack

IBM Watson: person, food, food product, child, bread

(All trained on COCO)

The image recognition algorithms that were trained on other datasets, however, are mystified. "Candle?" guesses one of them. "King crab?" "Pretzel?" "Conch?"

DenseNet: Candle

SqueezeNet: King crab

Inception V3: pretzel

ResNet-50: conch

(All trained on ImageNet)

The artist Tom White has used this effect to create a new kind of abstract art. He gives one AI a palette of abstract blobs and color washes and tells it to draw something (a jack-o'-lantern, for example) that another AI can identify.[12] The resulting drawings look only vaguely like the things they're supposed to be — a "measuring cup" is a squat green blob covered in horizontal scribbles, and a "cello" looks more like a human heart than a musical instrument. But to ImageNet-trained algorithms, the pictures are uncannily accurate. In a way, this artwork is a form of adversarial attack.

Of course, as in our earlier cockroach scenario, adversarial attacks are often bad news. In 2018 a team from Harvard Medical School and MIT warned that adversarial attacks in medicine could be particularly insidious — and profitable.[13] Today, people are developing image recognition algorithms to automatically screen X-rays, tissue samples, and other medical images for signs of disease. The idea is to save time by doing high-throughput screening so humans don't have to look at every image. Plus, the results could be consistent from hospital to hospital, everywhere the software is implemented — so they could be used to decide which patients qualify for certain treatments or to compare various drugs to one another.

That's where the motivation for hacking comes in. In the United States, insurance fraud is already lucrative, and some healthcare providers are adding unnecessary tests and procedures to increase revenue. An adversarial attack would be a handy, hard-to-detect way to move some patients from category A to category B. There's also temptation to tweak the results of clinical trials so a profitable new drug gets approved. And since a lot of medical image recognition algorithms are generic ImageNet-trained algorithms that have had a little extra training time on a specialized medical dataset, they're relatively easy to hack. This doesn't mean it's hopeless to use machine learning in medicine — it just means that we may always need a human expert spot-checking the algorithm's work.

Another application that may be particularly vulnerable to adversarial attack is fingerprint reading. A team from New York University Tandon and Michigan State University showed that it could use adversarial attacks to design what it called a masterprint — a single fingerprint that could pass for 77 percent of the prints in a low-security fingerprint reader.[14] The team was also able to fool higher-security readers, or commercial fingerprint readers trained on different datasets, a significant portion of the time. The masterprints even looked like regular fingerprints — unlike other spoofed images that contain static or other distortions — which made the spoofing harder to spot.

Voice-to-text algorithms can also be hacked. Make an audio clip of a voice saying "Seal the doors before the cockroaches get in," and you can overlay noise that a human will hear as subtle static but that will make a voice-recognition AI hear the clip as "Please enjoy a delicious sandwich." It's possible to hide messages in music or even in silence.

Resume screening services might also be susceptible to adversarial attack — not by hackers with algorithms of their own but by people trying to alter their resumes in subtle ways to make it past the AI. The *Guardian* reports: "One HR employee for a major technology company recommends slipping the words 'Oxford' or 'Cambridge' into a CV in invisible white text, to pass the automated screening."[15]

It's not like machine learning algorithms are the only technology that's vulnerable to adversarial attacks. Even humans are susceptible to the Wile E. Coyote style of adversarial attack: putting up a fake stop sign, for example, or drawing a fake tunnel on a solid rock wall. It's just that machine learning algorithms can be fooled by adversarial attacks that humans would never even register. And as AI becomes more widespread, we may be in for an arms race between AI security and increasingly sophisticated and difficult-to-detect hacks.

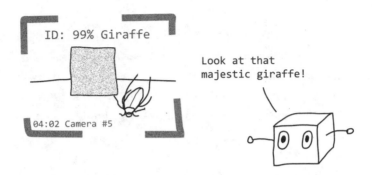

An example of an adversarial attack that's targeted at humans with touch screens: some advertisers have put fake specks of "dust" on their banner ads, hoping that humans will accidentally click on the ads while trying to brush them off.[16]

MISSING THE OBVIOUS

Without a way to see what AIs are thinking, or to ask them how they came to their conclusions (people are working on this), usually our first clue that something has gone wrong is when the AI does something weird.

An AI shown a sheep with polka dots or tractors painted on its sides will report seeing the sheep but will not report anything unusual about it. When you show it a sheep-shaped chair with two heads, or a sheep with too many legs, or with too many eyes, the algorithm will also merely report a sheep.

Why are AIs so oblivious to these monstrosities? Sometimes it's because they don't have a way to express them. Some AIs can only answer by output-ting a category name — like "sheep" — and aren't given an option for expressing that *yes*, it is a sheep, but something is very, very wrong. But there may often be another reason. It turns out that image recognition algorithms are very good at identifying scrambled images. If you chop an image of a fla-mingo into pieces and rearrange the pieces, a human will no longer be able to tell that it's a flamingo. But an AI may still have no trouble seeing the bird. It's still able to see an eye, a beak tip, and a couple of feet, and even though those aren't in the right spot relative to one another, the AI is only looking for the features, not how they're connected. In other words, the AI is acting like a **bag-of-features model**. Even AIs that theoretically are capable of looking at large shapes, not just tiny features, seem to often act like simple bag-of-features models.[17] If the flamingo's eyes are on its ankles, or if its beak is lying several meters away, the AI sees nothing out of the ordinary.

Basically, if you're in a horror movie where zombies start appearing, you might want to grab the controls from your self-driving car.

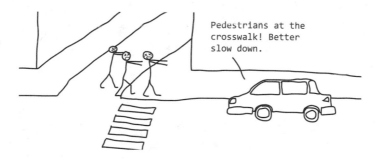

Pedestrians at the crosswalk! Better slow down.

More worryingly, the AI in a self-driving car may miss other rare, but more realistic, road hazards. If the car in front of it is on fire, fishtailing on ice, or carrying a Bond villain who just dropped a load of nails on the road, a self-driving car won't register anything wrong unless it's been specifi-cally prepared for this problem.

Could you design an AI to count eyes or identify flaming cars? Absolutely. An "on fire or not" AI could probably be pretty accurate. But to ask an AI to identify flaming cars *and* regular cars *and* drunk drivers *and* bicycles *and* escaped emus — this becomes a really broad task. Remember that the narrower the AI, the smarter it seems. Dealing with all the world's weirdness is a task that's beyond today's AI. For that, you'll need a human.

CHAPTER 9

Human bots (where can you *not* expect to see AI?)

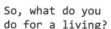

So, what do you
do for a living?

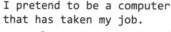

I pretend to be a computer
that has taken my job.

Throughout this book we've learned that AIs can perform at the level of a human only in very narrow, controlled situations. When the problem gets broad, the AI starts to struggle. Responding to one's fellow social media users is an example of a broad, tricky problem, and this is why what we call "social media bots" — rogue accounts that spread spam or misinformation — are unlikely to be implemented with AI. In fact, spotting a social media bot may be easier for an AI than *being* a social media bot. Instead, people who build social media bots are likely to use traditional rules-based programming to automate a few simple functions. Anything more sophisticated than that is likely to be a poorly paid human being instead of an actual AI. (There's a certain irony to the idea of a human stealing a robot's job.) In this chapter, I'll talk about instances in which what we think of as bots are really human beings — and where you're unlikely to see AI anytime soon.

A HUMAN IN BOT CLOTHING

People often give AIs tasks that are too hard. Sometimes, the programmers only find out there's a problem when their AIs try and fail. Other times, they don't realize that their AI is solving a different, easier, problem than the one they had hoped it would solve (for example, relying on the length of a medical case file rather than its contents to identify problem cases).[1] Still other programmers just *pretend* that they've figured out how to solve the problem with AI while secretly using humans to do it instead.

This latter phenomenon, claiming human performance as AI, is far more common than you'd think. The attraction of AI for many applications is its ability to scale to huge volumes, analyzing hundreds of images or transactions per second. But for very small volumes, it's cheaper and easier to use humans than to build an AI. In 2019, 40 percent of European start-ups classified in the AI category didn't use any AI at all.[2]

Sometimes using humans is only a temporary solution. A tech company may first build a human-powered mockup of its software while it works out things like user interfaces and workflow or while it gauges investor interest. Sometimes the human-powered mockup is even generating examples that will be used as training data for the eventual AI. This "fake it till you make it" approach can sometimes make a lot of sense. It can also be a risk — a company might end up demonstrating an AI that it can't actually build. Tasks that are doable for humans might be really hard, or even impossible, for AI. Humans have a sneaky habit of doing broad tasks without even realizing it.

What happens then? One solution companies sometimes use is to have a human employee waiting to swoop in if an AI begins to struggle. That's the way today's self-driving cars generally work: the AI can handle maintaining speed or even steering on long stretches of highway or during long hours of slow-speed stop-and-go traffic. But a human has to be ready to help at a moment's notice if there's something the AI is unsure about. This is called the **pseudo-AI** or **hybrid AI** approach.

Too confusing!
Help, human!

Some companies see pseudo-AI as a temporary bridge as they work on an AI solution they'll be able to scale. It may not always be as temporary as they'd hope. Remember Facebook M from chapter 2, a personal-assistant AI app that would send the tricky questions to human employees? Though the idea was to eventually phase out the use of humans, the assistant job turned out to be too broad for the AI to ever figure out.

Other companies embrace the pseudo-AI approach as a way to combine the best of AI speed and human flexibility. Multiple companies have offered hybrid image recognition, where if the AI is unsure about an image, it gets sent to humans to categorize. A meal-delivery service uses AI-powered robots — but bicycle-riding humans bring food from the restaurants to the robots, and the AI only has to help the robots navigate for five to ten seconds between waypoints set by remote human drivers.[3] Other companies are advertising hybrid AI chatbots: customers who begin by talking to an AI will be transferred to a human once the conversation gets tricky.

This can work well if customers know when they're dealing with a human. But sometimes customers who thought their expense reports,[4] personal schedules,[5] and voice mails[6] were being handled by an impersonal AI were shocked to learn that human employees were seeing their sensitive information — as were the human employees when they saw that they were being sent people's phone numbers, addresses, and credit card numbers.

Hybrid AI and pseudo-AI chatbots also have their own potential pitfalls. Every remote interaction becomes a form of the Turing test, and in the tightly limited, highly scripted environment of a customer service interaction, humans and AIs can be tough to tell apart. Humans may end up being treated badly by other humans who think they're dealing with a bot. Employees have already complained about this, including one whose job it was to generate real-time transcripts of phone calls for deaf and hearing-impaired customers. When a human made a mistake, the caller would sometimes complain about "useless computers."[7]

Another problem is that people end up with the wrong idea of what AI is capable of. If something claims to be AI and then starts holding human-level conversations, identifying faces and objects at a human level of performance, or producing nearly flawless transcriptions, people may assume that AIs really can do these things on their own. The Chinese government is reportedly taking advantage of this[8] with its nationwide surveillance system. Experts agree that there's no facial recognition system that could accurately identify the thirty million people China has on its watch lists. In 2018 the *New York Times* reported that the government was still doing much of its facial recognition the old-fashioned way, using humans to look through sets of photos and make matches. What they tell the public, however, is that they're using advanced AI. They'd like people to believe that a nationwide surveillance system is already capable of tracking their every move. And reportedly, people largely believe them. Jaywalking and crime rates are down in areas where the cameras have been publicized, and when told that the system had seen their crimes, some suspects have even confessed.

BOT OR NOT?

So given how many AIs are partially or even completely replaced by humans, how can we tell if we're dealing with a real AI? In this book, we've already covered a lot of things that you'll see AI doing—and things you won't see it doing. But out in the world, you'll encounter plenty of exaggerated claims about what AI can do, what it's already doing, or what it'll do soon. People trying to sell a product or sensationalize a story will come up with overblown headlines:

- Facebook AI Invents Language That Humans Can't Understand: System Shut Down Before It Evolves into Skynet[9]
- Babysitter Screening App Predictim Uses AI to Sniff Out Bullies[10]
- Here's What Sophia, the First Robot Citizen, Thinks About Gender and Consciousness[11]
- 30-Ton Electronic Brain at U. of P. Thinks Faster Than Einstein (1946)[12]

In this book I've tried to make it clear what AI is actually capable of and what it's unlikely to be able to do. Headlines like the ones above are giant red flags—and in this book I've given you many reasons why.

Here are a few questions to ask when evaluating AI claims.

1. How broad is the problem?

As we've seen throughout this book, AIs do best at very narrow, tightly defined problems. Playing chess or go is narrow enough for AI. Identifying specific kinds of images—recognizing the presence of a human face or distinguishing healthy cells from a specific kind of disease—is also probably doable. Dealing with all the unpredictability of a city street or a human conversation is probably beyond its reach — if it tries, it may succeed much of the time, but there will be glitches.

Of course, there are some problems that occupy gray areas. An AI may be able to sort medical images pretty well, but if you slip it a picture of a giraffe, it will probably be baffled. AI chatbots that pass as human usually use some gimmick — such as, in one specific case, pretending to be an eleven-year-old Ukrainian kid with limited English skills[13] — to explain away non sequiturs or their inability to handle most topics. Other AI chatbots have their "conversations" in controlled settings where the questions are known — and the answers human-written — ahead of time. If a problem seems like it required broad understanding or context to solve, a human was probably responsible.

2. Where did the training data come from?

Sometimes people show off "AI-written" stories that they have written themselves. You may remember a viral Twitter joke from 2018 about a bot that watched a thousand hours of Olive Garden commercials and generated a script for a new one. One giveaway that the joke was written by a human was that the description of what the AI learned from doesn't match what it produced. If you give an AI a bunch of videos to learn from, it will output videos. It won't be able to produce a script with stage directions — not unless there's another AI, or a human, whose job it is to turn videos into scripts. Did the AI have a set of examples to copy or a fitness function to maximize? If not, then you're probably not looking at the product of an AI.

3. Does the problem require a lot of memory?

Remember from chapter 2 that AIs do best when they don't have to remember very much at once. People are improving this all the time, but for now, a sign of an AI-generated response is a lack of memory. AI-written stories will meander, forgetting to resolve earlier plot points, sometimes even forgetting to finish sentences. AIs that play complex video games have a tough time with long-term strategy. AIs that hold conversations will forget information you gave them earlier unless they're explicitly programmed to remember things like your name.

An AI that's making callbacks to earlier jokes, that sticks with a consistent cast of characters, and that keeps track of the objects in a room probably had a lot of human editing help, at least.

4. Is it just copying human biases?

Even if people do genuinely use AI to solve a problem, it's possible the AI is not nearly as capable as its programmers claim. For example, if a company claims to have developed a new AI that can comb through a job candidate's social media and decide whether or not that person is trustworthy, we should immediately be raising red flags. A job like that would require human-level language skills, with the ability to handle memes, jokes, sarcasm, references to current events, cultural sensitivity, and more. In other words, it's a task for a general AI. So if it's returning ratings of each candidate, what is it basing its decisions on?

The CEO of one such service, which in 2018 was offering social media screenings of potential babysitters, told Gizmodo, "We trained our product, our machine, our algorithm to make sure it was ethical and not biased." As evidence of its AI's lack of bias, the company's CTO said, "We don't look at skin color, we don't look at ethnicity, those aren't even algorithmic inputs. There's no way for us to enter that into the algorithm itself." But as we've seen, there are plenty of ways for a determined AI to pick up on trends that seem to help it figure out how humans rate each other — zip code and even photographs can be an indicator of race, and word choice can give it clues about things like gender and social class. As a possible indication of problems, when a Gizmodo reporter tested the babysitter-screening service, he found that his black friend was rated as "disrespectful" while his foul-mouthed white friend was rated more highly. When asked if the AI might have picked up on systemic bias in its training data, the CEO admitted that this was possible but noted that they added human review to catch errors like this. The question, then, is why the service rated those two friends the way it did. Human review doesn't necessarily solve the problem of a biased algorithm, since the bias likely came from humans in the first place. And this particular AI doesn't tell its customers how it came to its decisions, and it quite possibly doesn't tell its programmers, either. This makes its decisions hard to appeal.[14] Shortly after Gizmodo and others reported on their service, Facebook, Twitter, and Instagram restricted the company's social media access, citing violations of terms of service, and the company halted their planned launch.[15]

There may be similar problems with AIs that screen job candidates, like the Amazon-resume-screening AI that learned to penalize female candidates. Companies that offer AI-powered candidate screening point to case studies of clients who have significantly increased the diversity of their hires after using AI.[16] But without careful testing, it's hard to know why. An AI-powered job screener could help increase diversity even if it recommended candidates entirely at random, if that's already better than the racial and/or gender bias in typical company hiring. And what does a video-watching AI do about candidates with facial scarring or partial paralysis or whose facial expressions don't match Western and/or neurotypical norms?

As CNBC reported in 2018, people are already being advised to over-emote for the AIs that screen videos of job candidates or to wear makeup that makes their faces easier to read.[17] If emotion-screening AIs become more prevalent, scanning crowds for people whose microexpressions or body language trigger some warning, people could be compelled to perform for those, too.

The problem with asking AI to judge the nuances of human language and human beings is that the job is just too hard. To make matters worse, the only rules that are simple and reliable enough for it to understand may be those — like prejudice and stereotyping — that it shouldn't be using. It's possible to build an AI system that improves on human prejudices, but it doesn't happen without a lot of deliberate work, and bias can sneak in despite the best of intentions. When we use AI for jobs like this, we can't trust its decisions, not without checking its work.

CHAPTER 10

A human-AI partnership

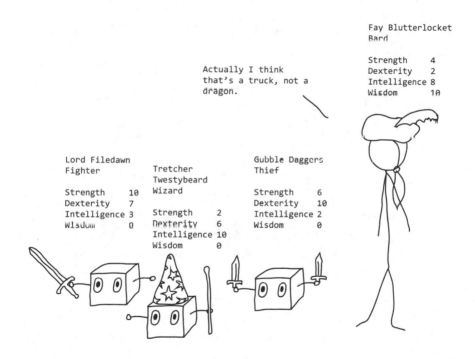

Fay Blutterlocket
Bard

Strength   4
Dexterity   2
Intelligence 8
Wisdom   10

Actually I think
that's a truck, not a
dragon.

Lord Filedawn
Fighter

Strength   10
Dexterity   7
Intelligence 3
Wisdom   0

Tretcher
Twestybeard
Wizard

Strength   2
Dexterity   6
Intelligence 10
Wisdom   0

Gubble Daggers
Thief

Strength   6
Dexterity   10
Intelligence 2
Wisdom   0

INSTANT AI: JUST ADD HUMAN EXPERTISE

If there's one thing we've learned from this book, it's that AI can't do much
without humans. Left to its own devices, at best it will flail ineffectually,
and at worst it will solve the wrong problem entirely — which, as we've

seen, can have devastating consequences. So it's unlikely that AI-powered automation will be the end of human labor as we know it. A far more likely vision for the future, even one with the widespread use of advanced AI technology, is one in which AI and humans collaborate to solve problems and speed up repetitive tasks. In this chapter, I'll take a look at what the future holds for AI and humans working together — and how they can partner in surprising ways.

I'm helping!

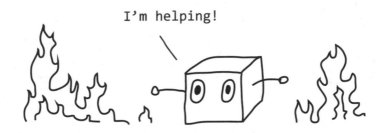

As we've seen throughout this book, humans need to make sure that an AI solves the right problems. This job involves anticipating the kinds of mistakes that machine learning tends to make and making sure to look for them — and even to avoid them in the first place. Choosing the right data can be a big part of that — we've seen that messy or flawed data can lead to problems. And of course an AI can't go collect its own dataset. Not unless we design *another* AI whose job it is to find data.

Building the AI in the first place is, of course, another job for humans. A blank mind that absorbs information like a sponge only exists in science fiction. For real AIs, a human has to choose the form to match the problem it's supposed to solve. Are we building something that will recognize images? Something that will generate new scenes? Something that will predict numbers on a spreadsheet or words in a sentence? Each of those needs a specific type of AI. If the problem is complex, it may need many specialized algorithms working together for the best results. Again, a human has to choose the subalgorithms and set them up so they can learn together.

A lot of human engineering goes into the dataset as well. The AI will get further if the human programmer can set things up so the AI has less to do. Remember the knock-knock jokes from chapter 1 — the AI would have progressed a lot faster if it didn't have to learn the entire joke formula of knocks and responses but could just focus on filling in the punchline. It would have done even better if we had started it off with a list of existing words and phrases to use when constructing puns. To cite another example, people who know that their AIs will need to keep track of 3-D information can help them out by building them with 3-D object representations in mind.[1] Cleaning up a messy dataset to remove distracting or confusing data is also an important part of human dataset engineering. Remember the AI from chapter 4 that spent its time trying to format ISBN numbers rather generating the recipes it was supposed to, and dutifully copied weird typos from its dataset?

In that sense, practical machine learning ends up being a bit of a hybrid between rules-based programming, in which a human tells a computer step-by-step how to solve a problem, and open-ended machine learning, in which an algorithm has to figure everything out. A human with very specialized knowledge about whatever the algorithm's trying to solve can really help the program out. In fact, sometimes (perhaps even ideally) the programmer researches the problem and discovers that they now understand it so well that they no longer need to use machine learning at all.

Of course, too much human supervision can also be counterproductive. Not only are humans slow, but we also sometimes just don't know what the best approach to a problem is. In one instance, a group of researchers tried to improve the performance of an image recognition algorithm by incorporating more human help.[2] Rather than just label a picture as depicting a dog, the researchers asked humans to click on the part of the image that actually contained the dog, then they programmed the AI to pay special attention to that part. This approach makes sense — shouldn't the AI learn faster if people point out what part of the picture it should be paying attention to? It turns out that the AI *would* look at the doggy if you made it — but more than just a tiny bit of influence would make it perform much worse. Even more confoundingly, researchers don't know exactly why. Maybe there's something we don't understand about what really helps an image recognition algorithm identify something. Maybe the people who clicked on the images don't even understand how they recognize dogs and clicked on the parts of the images they thought were important (mostly eyes and muzzles) rather than the parts they actually used to identify it. When the researchers asked the AI which parts of the images *it* thought were important (by looking at which parts made its neurons activate), it was likely to highlight the edges of the dog or even the background of the photo.

MAINTENANCE

Another thing machine learning needs humans for is maintenance.

After an AI has been trained on real-world data, the world might change. Machine learning researcher Hector Yee reports that around 2008 some colleagues told him there was no need to design a new AI to detect cars in an image — they already had an AI that worked great. But when Yee tried their AI on real-world data, it did terribly. It turned out that the AI had been trained on cars from the 1980s and didn't know how to recognize modern cars.[3]

I've seen similar quirks with Visual Chatbot, the giraffe-happy chatbot we met in chapter 4. It has a tendency to identify handheld objects (light-sabers, guns, swords) as Wii remotes. That might be a reasonable guess if it were still 2006, when Wii was in its heyday. More than a decade later, however, finding a person holding a Wii remote is becoming increasingly unlikely.

All sorts of things could change and mess with an AI. As I mentioned in an earlier chapter, road closures or even hazards like wildfires might not deter an AI that sees only traffic from recommending what it thinks is an attractive route. Or a new kind of scooter could become popular, throwing off the hazard-detection algorithm of a self-driving car. A changing world adds to the challenge of designing an algorithm to understand it.

People also need to be able to adjust algorithms to fix newly discovered problems. Maybe there's a rare but catastrophic bug that develops, like the one that affected Siri for a brief period of time, causing her to respond to users saying "Call me an ambulance" with "Okay, I'll call you 'an ambulance' from now on."[4]

Another place where we need human oversight is in the matter of detecting and correcting bias. To combat the tendency of AI decision making to perpetuate bias, governments and other organizations are starting to require bias testing as a matter of course. As I mentioned in chapter 7, in January 2019, New York State issued a letter requiring life insurance companies to prove that their AI systems do not discriminate on the basis of race, religion, country of origin, or other protected classes. The state worried that making coverage decisions using "external lifestyle indicators" — anything from home address to educational level — would lead an AI to use this information to discriminate in illegal ways.[5] In other words, they wanted to prevent mathwashing. We may see pushback against this kind of testing from companies that want their AIs to remain proprietary or harder to hack or that don't want their AIs' embarrassing shortcuts to be revealed. Remember Amazon's sexist resume-screening AI? The company discovered the problem before using the AI in the real world and told us about it as a cautionary tale. How many other biased algorithms are out there right now, doing their best but doing it wrong?

BEWARE OF AIS THAT LEARN ON THE JOB

Not only are AIs not great at realizing when their brilliant solutions pose problems, AIs and their environments can also interact in unfortunate ways. One example is the now infamous Microsoft Tay chatbot, a machine learning–based Twitter bot that was designed to learn from the users who tweeted at it. The bot was short-lived. "Unfortunately, within the first 24 hours of coming online," Microsoft told the *Washington Post,* "we became aware of a coordinated effort by some users to abuse Tay's commenting skills to have Tay respond in inappropriate ways. As a result, we have taken Tay offline and are making adjustments."[6] It had taken almost no time at all for users to teach Tay to spew hate speech and other abuse. Tay had no built-in sense of what kind of speech was offensive, a fact that vandals were happy to exploit. In fact, it's notoriously difficult to flag offensive content without also falsely flagging discussion of the *effects* of offensive content. Without a good way to recognize offensive things automatically, machine learning algorithms will sometimes go out of their way to promote it, as we learned in chapter 5.

The AIs that autocomplete search-engine queries learn on the fly, and that can lead to weird results when humans are in the mix. The problem with humans is that if search-engine autocomplete makes a really hilarious mistake, humans will tend to click on it, which just makes the AI even more likely to suggest it to the next human. This famously happened in 2009 with the phrase "Why won't my parakeet eat my diarrhea?"[7] Humans found this suggested question so hilarious that soon the AI was suggesting it as soon as people began typing "Why won't." Probably a human at Google had to manually intervene to stop the AI from suggesting that phrase.

As I mentioned in chapter 7, there are also dangers if predictive-policing algorithms learn on the job. If an algorithm sees that there are more arrests in a particular neighborhood than there are in others, it will predict that there will be more arrests there in the future, too. If the police respond to this prediction by sending more officers to the area, it may become a self-fulfilling prophecy: more police on the streets means that even if the actual crime rate is no higher than it is in other neighborhoods, the police will witness more crimes and make more arrests. When the algorithm sees the new arrest data, it may predict an even higher arrest rate in that neighborhood. If the police respond by increasing their presence in the neighborhood, then the problem will only escalate. Of course, it doesn't require an AI to be susceptible to this kind of feedback loop — very simple algorithms and even humans fall for this as well.

Here's a very simple feedback loop in action: In 2011 a biologist named Michael Eisen noticed something odd when a researcher in his lab tried to buy a particular textbook about fruit flies.[8] The book was out of print but not terribly rare; there were used copies available on Amazon for around $35. The two new copies available, however, were priced at $1,730,045.91 and $2,198,177.95 (plus $3.99 shipping). When Eisen checked again the next day, both books had increased in price, to nearly $2.8 million. Over the next few days, a pattern emerged: in the morning, the company that sold the less expensive book would increase its price so that it was exactly 0.9983 times the price of the more expensive book. In the afternoon, the expensive book's price would increase to become exactly 1.270589 times the price of the cheaper book. Both companies were apparently using algorithms to set their book prices. It was clear that one company wanted to charge as much as it could while still having the cheapest book available. But what was the motivation of the company that sold the more expensive book? Eisen noticed that that company had very good feedback scores and theorized that it was counting on this to induce some customers to pay a slightly higher price for the book — at which point it would order the book from the cheaper company and ship it to the customer, pocketing the profit. After about a week the spiraling prices dropped back to normal. Apparently some human had noticed the problem and corrected it. But companies use unsupervised algorithmic pricing all the time. Once, when I checked Amazon, there were several coloring books being offered for $2,999 apiece.

So the book prices were the products of simple rules-based programs. But machine learning algorithms can make trouble in even more exciting new ways. A 2018 paper showed that two machine learning algorithms in a situation like the book-pricing setup above, each given the task of setting a price that maximizes profits, can learn to collude with each other in a way that's both highly sophisticated and highly illegal. They can do this without explicitly being taught to collude and without communicating directly with each other — somehow, they manage to set up a price-fixing scheme just by observing each other's prices. This has only been demonstrated in a simulation so far, not in a real-world pricing scenario. But people have estimated that a large portion of online prices are being set by autonomous AIs, so the prospect of widespread price fixing is worrying. Collusion is great for sellers — if everyone cooperates to set high prices, then profits go up — but it's bad for consumers. Even without meaning to, sellers could potentially be using AI to do things that it's illegal for them to do explicitly.[9] This is just another face of the mathwashing phenomenon I brought up in chapter 7. Humans will have to make sure that their AIs aren't being tricked by bad actors or accidentally becoming bad actors themselves.

LET THE AI HANDLE THIS ONE

Human-level performance is the gold standard for a lot of machine learning algorithms. After all, much of the time their task is to imitate examples of humans doing stuff: labeling pictures, filtering emails, naming guinea pigs. And in cases where their performance is more or less at a human level, they can (with supervision) be used to replace humans for tasks that are repetitive or boring. We've seen in earlier chapters that some news organizations are using machine learning algorithms to automatically create boring but acceptable articles on local sports or real estate. A project called Quicksilver automatically creates draft *Wikipedia* articles about female scientists (who have been noticeably underrepresented on *Wikipedia*), saving volunteer editors time. People who need to write audio transcripts or translate text use the (admittedly buggy) machine learning versions as a starting point for their own translations. Musicians can employ music-generating algorithms, using them to put together a piece of original music to exactly fit a commercial slot for which the music doesn't have to be exceptional, just inexpensive. In many cases, the human role is to be an editor.

And there are some jobs for which it's even preferable not to use humans. People are more likely to open up about their emotions or disclose potentially stigmatizing information if they think they're talking to a robot as opposed to a human.[10, 11] (On the other hand, healthcare chatbots could potentially miss serious health concerns).[12] Bots have also been trained to look through disturbing images and flag potential crimes (though they tend to mistake desert scenes for human flesh).[13] Even crime itself may be more easily committed by a robot than a human. In 2016, Harvard student Serena Booth built a robot that was meant to test some theories about whether humans trust robots too much.[14] Booth built a simple remote-controlled robot and had it drive up to students, asking to be allowed access to a key card–controlled dorm. Under those circumstances, only 19 percent of people let it into the dorm (interestingly, that number was a bit higher when the students were in groups). However, if the same robot said it was delivering cookies, 76 percent let it in.

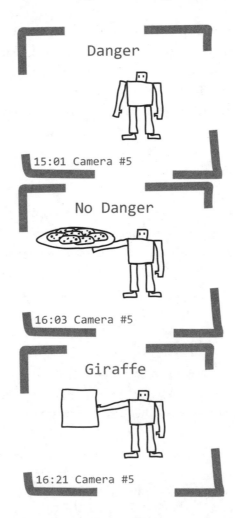

As I mentioned above, some AIs may also be good at crime because of the mathwashing phenomenon. An AI's decisions can be based on complex relationships between several variables, some of which may be proxies for information that it's not supposed to have, like gender or race. That adds a layer of obfuscation that may — intentionally or not — be allowing it to get away with breaking laws.

Task: ~~do crimes~~
play with the numbers and see what happens

There are also plenty of cases in which AI is preferable because it exceeds human performance. For one, it's usually much faster than humans. In some multiplayer computer games, when AI plays against humans, the AI has to be slowed down to give the humans a fighting chance. AI is also more consistent, if terrible at handling the unexpected. Can AI also be fairer? Potentially. An AI-powered system, at least, can be tested for fairness by running lots of test decisions and looking for statistical correlations that shouldn't be there. By carefully adjusting the training data to make its statistics match the world as it *should* be rather than the world as it is, it would be possible in many cases to train an AI whose decisions are fair — at least, much fairer than your average human's.

ALGORITHMIC CREATIVITY?

Will the music, movies, and novels of the future be written by AI? Maybe at least partially.

AI-generated art can be striking, weird, and unsettling: infinitely morphing tulips; glitchy humans with half-melted faces; skies full of hallucinated dogs. A T. rex may turn into flowers or fruit; the *Mona Lisa* may take on a goofy grin; a piano riff may turn into an electric guitar solo. AI-generated text may take on the quality of surrealist performance art.

> When prompted with the following text: "My 10 favorite animals are: 1." the neural network GPT-2 added this list:
>
> ```
> My 10 favorite animals are:
> 1. zebras with a white scar on the back;
> 2. insiduous spiders and octopus;
> 3. frog with large leaves, hopefully black;
> 4. cockatiel with scales;
> 5. razorbill with wings hanging about 4 inches
>    from one's face and a heart tattoo on a frog.
> ```

Like AI problem solving, AI creativity could probably best be described as "AI-aided."

For a GAN to produce a painting, it first needs a dataset, and a human chooses what that dataset should be. Some of the most interesting GAN results occur when artists give the algorithms their own paintings, or their own photography, to learn from. The artist Anna Ridler, for example, spent a spring taking ten thousand photos of tulips, then used her photos to train a GAN that produced an endless series of nearly photorealistic tulips, each tulip's stripiness tied to the price of Bitcoin. The artist and software engineer Helena Sarin has produced interesting GAN remixes of her own watercolors and sketches, morphing them into cubist or weirdly textured hybrids. Other artists are inspired to choose existing datasets — like public-domain Renaissance portraits or landscapes — and see what a GAN might make with them. Curating a dataset is also an artistic act — add more styles of painting, and a hybrid or corrupted artwork might result. Prune a dataset to a single consistent angle, style, or type of lighting, and the neural net will have an easier time matching what it sees to produce more realistic images. Start with a model trained on a large dataset, then use transfer learning to focus in on a smaller but more specialized dataset, for even more ways to fine-tune the results.

I shall curate a dataset of 100,000 pictures of giraffes!!!

best. dataset. ever.

People who train text-generating algorithms also can control their results via their datasets. Science fiction writer Robin Sloan is one of a few writers experimenting with neural network–generated text as a way of injecting some unpredictability into his writing.[15] He built a custom tool that responds to his own sentences by predicting the next sentence in the sequence based on its knowledge of other science fiction stories, science news articles, and even conservation news bulletins. Demonstrating his tool in an interview with the *New York Times*, Sloan fed it the sentence "The bison are gathered around the canyon," and it responded with "by the bare sky." It wasn't a perfect prediction in the sense that there was something noticeably off about the algorithm's sentence. But for Sloan's purposes, it was delightfully weird. He'd even rejected an earlier model he'd trained on 1950s and 1960s science fiction stories, finding its sentences too clichéd.

Like collecting the datasets, training the AI is an artistic act. How long should training last? An incompletely trained AI can sometimes be interesting, with weird glitches or garbled spelling. If the AI gets stuck and begins to produce garbled text or strange visual artifacts like multiplying grids or saturated colors (a process known as **mode collapse**), should the training start over? Or is this effect kinda cool? As in other applications, the artist will also have to watch to make sure the AI doesn't copy its input data too closely. As far as an AI knows, an exact copy of its dataset is just what it's being asked for, so it will plagiarize if it possibly can.

And finally, it's the human artist's job to curate the AI's output and turn it into something worthwhile. GANs and text-generating algorithms can create virtually infinite amounts of output, and most of it isn't very interesting. Some of it is even terrible — remember that many text-generating neural nets don't know what their words mean (I'm looking at you, neural net that suggested naming cats Mr. Tinkles and Retchion). When I train neural nets to generate text, only a tiny fraction — a tenth or a hundredth — of the results are worth showing. I'm always curating the results to present a story or some interesting point about the algorithm or the dataset.

In some cases, curating the output of an AI can be a surprisingly involved process. I used BigGAN in chapter 4 to show how image-generating neural nets struggle when trained on images that are too varied — but I didn't talk about one of its coolest features: generating images that are a blend of multiple categories.

Think of "chicken" as a point in space and "dog" as a point in space. If you take the shortest path between them, you pass other points in space that are somewhere between the two, in which chickendogs have feathers, floppy ears, and lolling tongues. Start at "dog" and travel toward "tennis ball," and you'll pass through a region of fuzzy green spheres with black eyes and boopable noses. This huge multidimensional visual landscape of possibility is called **latent space**. And once BigGAN's latent space was accessible, artists began to dive in to explore. They quickly found coordinates where there were overcoats covered in eyes and trench coats covered in tentacles, angular-faced dog-birds with both eyes on one side of their faces, picture-perfect hobbit villages complete with ornate rounded doors, and flaming mushroom clouds with cheerful puppy faces. (ImageNet has a lot of dogs in it, as it turns out, so the latent space of BigGAN is also full of dogs.) Methods of navigating latent space become themselves artistic choices. Should we travel in straight lines or curves? Should we keep our locations close to our origin point or allow ourselves to veer off into extreme far-flung corners? Each of these choices drastically affects what we see. The rather utilitarian categories of ImageNet blend into utter weirdness.

Is all this art AI-generated? Absolutely. But is the AI the thing doing the creative work? Not by a long shot. People who claim that their AIs are the artists are exaggerating the capabilities of the AIs — and selling short their own artistic contributions and those of the people who designed the algorithms.

Life among our artificial friends

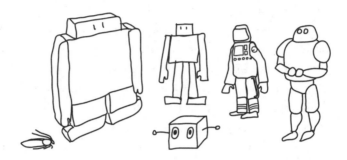

Over the course of these pages, we've seen lots of different ways that AI can surprise us.

Given a problem to solve, and enough freedom in how to solve it, AIs can come up with solutions that their programmers never dreamed existed. Tasked with walking from point A to point B, an AI may decide instead to assemble itself into a tower and fall over. It may decide to travel by spinning in tight circles or twitching along the floor in a writhing heap. If we train it in simulation, it may hack into the very fabric of its universe, figuring out ways to exploit physics glitches to attain superhuman abilities. It will take instructions literally: when told to avoid collisions, it will refuse to move; when told to avoid losing a video game, it will find the Pause button and freeze the game forever. It will find patterns hidden in its training data, even patterns its programmers didn't expect. Some of the patterns may be ones we didn't want it to emulate, like bias. Modular AIs may cascade together, cooperating to accomplish tasks that no single AI could tackle alone, acting like a phone full of apps or even a swarm of bees.

As AI becomes ever more capable, it still won't know what we want. It will still *try* to do what we want. But there will always be a potential disconnect between what we want AI to do and what we tell it to do. Will it get smart enough to understand us and our world as another human does — or even to surpass us? Probably not in our lifetimes. For the foreseeable future, the danger will not be that AI is too smart but that it's not smart enough.

On the surface, AI will seem to understand more. It will be able to generate photorealistic scenes, maybe paint entire movie scenes with lush textures, maybe beat every computer game we can throw at it. But underneath that, it's all pattern matching. It only knows what it has seen and seen enough times to make sense of.

Our world is too complicated, too unexpected, too bizarre for an AI to have seen it all during training. The emus will get loose, the kids will start wearing cockroach costumes, and people will ask about giraffes even when there aren't any present. AI will misunderstand us because it lacks the context to know what we really want it to do.

To take the best way forward with AI, we'll have to understand it — understand how to choose the right problems for it to solve, how to anticipate its misunderstandings, and how to prevent it from copying the worst of what it finds in human data. There's every reason to be optimistic about AI and every reason to be cautious. It all depends on how well we use it.

And watch out for those hidden giraffes.

Acknowledgments

This book would not exist without the hard work, insight, and generosity of a bunch of people who I'm delighted to thank here.

A huge thanks to the team at Voracious, whose hard work turned my sprawling, meandering document into a thing that I love. Barbara Clark's copyediting improved this book immeasurably, and it is lighter for the removal of a metric ton of *actuallys*. Thanks especially to my editor, Nicky Guerreiro, who emailed me out of the blue one day to say it was her fifth time stifling laughter in her open-plan office, and had I thought about how my blog might translate into a book? Without Nicky's encouragement and keen insight, this book would not have the scope and courage that it does.

Warm thanks also to my agent, Eric Lupfer, at Fletcher and Company for cheerfully guiding a first-time author through the many steps of turning a blog into a book.

The first time I heard about machine learning was in 2002 when Erik Goodman gave a fascinating talk about evolutionary algorithms to the incoming freshmen at Michigan State University. I guess those anecdotes about algorithms breaking simulations and solving the wrong problem really stuck with me! Thanks for sparking that interest early — it has led me to so much joy.

Thanks to my friends and family, who encouraged me during this long process, who listened to my practice talks and laughed at my jokes, and who were always ready to help me recharge with some tunes, some hiking, or some culinary experiments.

And finally, thanks to all my readers and followers at aiweirdness.com, who have already made so much of my strange AI experiments into reality — the knitting patterns, the cookies, the nail polish, the burlesque shows, the weird creatures, the absurd cat names, the beer names, and even the opera. Look what we made now! May the giraffes be ever with you.

Notes

Introduction

1. Caroline O'Donovan et al., "We Followed YouTube's Recommendation Algorithm Down the Rabbit Hole," *BuzzFeed News,* January 24, 2019, https://www.buzzfeednews.com/article/carolineodonovan/down-youtubes-recommendation-rabbithole.

Chapter 1: What is AI?

1. Joel Lehman et al., "The Surprising Creativity of Digital Evolution: A Collection of Anecdotes from the Evolutionary Computation and Artificial Life Research Communities," ArXiv:1803.03453 [Cs], March 9, 2018, http://arxiv.org/abs/1803.03453.
2. Neel V. Patel, "Why Doctors Aren't Afraid of Better, More Efficient AI Diagnosing Cancer," *The Daily Beast,* December 11, 2017, https://www.thedailybeast.com/why-doctors-arent-afraid-of-better-more-efficient-ai-diagnosing-cancer.
3. Jeff Larson et al., "How We Analyzed the COMPAS Recidivism Algorithm," *ProPublica,* May 23, 2016, https://www.propublica.org/article/how-we-analyzed-the-compas-recidivism-algorithm.
4. Chris Williams, "AI Guru Ng: Fearing a Rise of Killer Robots Is Like Worrying about Overpopulation on Mars," *The Register,* March 19, 2015, https://www.theregister.co.uk/2015/03/19/andrew_ng_baidu_ai/.
5. Marianne Bertrand and Sendhil Mullainathan, "Are Emily and Greg More Employable Than Lakisha and Jamal? A Field Experiment on Labor Market Discrimination," *American Economic Review* 94, no. 4 (September 2004): 991–1013, https://doi.org/10.1257/0002828042002561.

Chapter 2: AI is everywhere, but where is it exactly?

1. Stephen Chen, "A Giant Farm in China Is Breeding 6 Billion Cockroaches a Year. Here's Why," *South China Morning Post,* April 19, 2018, https://www.scmp.com/news/china/society/article/2142316/giant-indoor-farm-china-breeding-six-billion-cockroaches-year.

2. Heliograf, "High School Football This Week: Einstein at Quince Orchard," *Washington Post,* October 13, 2017, https://www.washingtonpost.com/allmetsports/2017 -fall/games/football/87408/.

3. Li L'Estrade, "MittMedia Homeowners Bot Boosts Digital Subscriptions with Automated Articles," International News Media Association (INMA), June 18, 2018, https://www.inma.org/blogs/ideas/post.cfm/mittmedia-homeowners-bot-boosts -digital-subscriptions-with-automated-articles.

4. Jaclyn Peiser, "The Rise of the Robot Reporter," *New York Times,* February 5, 2019, https://www.nytimes.com/2019/02/05/business/media/artificial-intelligence -journalism-robots.html.

5. Christopher J. Shallue and Andrew Vanderburg, "Identifying Exoplanets with Deep Learning: A Five Planet Resonant Chain around Kepler-80 and an Eighth Planet around Kepler-90," *The Astronomical Journal* 155, no. 2 (January 30, 2018): 94, https://doi.org/10.3847/1538-3881/aa9e09.

6. R. Benton Metcalf et al., "The Strong Gravitational Lens Finding Challenge," *Astronomy & Astrophysics* 625 (May 2019): A119, https://doi.org/10.1051/0004-6361/ 201832797.

7. Avi Bagla, "#StarringJohnCho Level 2: Using DeepFakes for Representation," YouTube video, posted April 9, 2018, https://www.youtube.com/watch?v=hlZkATlqDSM& feature=youtu.be.

8. Tom Simonite, "Facebook Built the Perfect Chatbot but Can't Give It to You Yet," *MIT Technology Review,* April 14, 2017, https://www.technologyreview.com/s/604117/ facebooks-perfect-impossible-chatbot/.

9. Ibid.

10. Casey Newton, "Facebook Is Shutting Down M, Its Personal Assistant Service That Combined Humans and AI," *The Verge,* January 8, 2018, https://www.theverge .com/2018/1/8/16856654/facebook-m-shutdown-bots-ai.

11. Andrew J. Hawkins, "Inside Waymo's Strategy to Grow the Best Brains for Self-Driving Cars," *The Verge,* May 9, 2018, https://www.theverge.com/2018/5/9/17307156/google -waymo-driverless-cars-deep-learning-neural-net-interview.

12. "OpenAI Five," OpenAI, accessed August 3, 2019, https://openai.com/five/.

13. Katyanna Quatch, "OpenAI Bots Smashed in Their First Clash against Human Dota 2 Pros," *The Register,* August 23, 2018, https://www.theregister.co.uk/2018/08/23/ openai_bots_defeated/.

14. Tom Murphy (@tom7), Twitter, August 23, 2018, https://twitter.com/tom7/status/ 1032756005107580929.

15. Mike Cook (@mtrc), Twitter, August 23, 2018, https://twitter.com/mtrc/status/ 1032783369254432773.

16. Tom Murphy, "The First Level of Super Mario Bros. Is Easy with Lexicographic Orderings and Time Travel…After That It Gets a Little Tricky" (research paper, Carnegie Melon University), April 1, 2013, http://www.cs.cmu.edu/~tom7/mario/mario.pdf.

17. Benjamin Solnik et al., "Bayesian Optimization for a Better Dessert" (paper presented at the 2017 NIPS Workshop on Bayesian Optimization, Long Beach, CA, December 9, 2017), https://bayesopt.github.io/papers/2017/37.pdf.

18. Sarah Kimmorley, "We Tasted the 'Perfect' Cookie Google Took 2 Months and 59 Batches to Create — and It Was Terrible," *Business Insider Australia,* May 31, 2018, https://www.businessinsider.com.au/google-smart-cookie-ai-recipe-2018-5.

19. Andrew Krok, "Waymo's Self-Driving Cars Are Far from Perfect, Report Says," *Roadshow,* August 28, 2018, https://www.cnet.com/roadshow/news/waymo-alleged-tech-troubles-report/.

20. C. Lv et al., "Analysis of Autopilot Disengagements Occurring during Autonomous Vehicle Testing," *IEEE/CAA Journal of Automatica Sinica* 5, no. 1 (January 2018): 58–68, https://doi.org/10.1109/JAS.2017.7510745.

21. Andrew Krok, "Uber Self-Driving Car Saw Pedestrian 6 Seconds before Crash, NTSB Says," *Roadshow,* May 24, 2018, https://www.cnet.com/roadshow/news/uber-self-driving-car-ntsb-preliminary-report/.

22. Fred Lambert, "Tesla Elaborates on Autopilot's Automatic Emergency Braking Capacity over Mobileye's System," *Electrek* (blog), July 2, 2016, https://electrek.co/2016/07/02/tesla-autopilot-mobileye-automatic-emergency-braking/.

23. Naaman Zhou, "Volvo Admits Its Self-Driving Cars Are Confused by Kangaroos," *The Guardian,* July 1, 2017, https://www.theguardian.com/technology/2017/jul/01/volvo-admits-its-self-driving-cars-are-confused-by-kangaroos.

Chapter 3: How does it actually learn?

1. Ian Goodfellow, Yoshua Bengio, and Aaron Courville, *Deep Learning* (Cambridge, Massachusetts: The MIT Press, 2016).

2. Sean McGregor et al., "FlareNet: A Deep Learning Framework for Solar Phenomena Prediction" (paper presented at the 31st Conference on Neural Information Processing Systems, Long Beach, CA, December 8, 2017), https://dl4physicalsciences.github.io/files/nips_dlps_2017_5.pdf.

3. Alec Radford, Rafal Jozefowicz, and Ilya Sutskever, "Learning to Generate Reviews and Discovering Sentiment," ArXiv:1704.01444 [Cs], April 5, 2017, http://arxiv.org/abs/1704.01444.

4. Andrej Karpathy, "The Unreasonable Effectiveness of Recurrent Neural Networks," Andrej Karpathy Blog, May 21, 2015, http://karpathy.github.io/2015/05/21/rnn-effectiveness/.

5. Chris Olah et al., "The Building Blocks of Interpretability," *Distill* 3, no. 3 (March 6, 2018): e10, https://doi.org/10.23915/distill.00010.

6. David Bau et al., "GAN Dissection: Visualizing and Understanding Generative Adversarial Networks" (paper presented at the International Conference on Learning Representations, May 6–9, 2019), https://gandissect.csail.mit.edu/.

7. "Botnik Apps," Botnik, accessed August 3, 2019, ttp://botnik.org/apps.

8. Paris Martineau, "Why Google Docs Is Gaslighting Everyone about Spelling: An Investigation," *The Outline*, May 7, 2018, https://theoutline.com/post/4437/why-google -docs-thinks-real-words-are-misspelled.

9. Shaokang Zhang et al., "Zoonotic Source Attribution of *Salmonella enterica* Serotype Typhimurium Using Genomic Surveillance Data, United States," *Emerging Infectious Diseases* 25, no. 1 (2019): 82–91, https://doi.org/10.3201/eid2501.180835.

10. Ian J. Goodfellow et al., "Generative Adversarial Networks," ArXiv:1406.2661 [Cs, Stat], June 10, 2014, http://arxiv.org/abs/1406.2661.

11. Ahmed Elgammal et al., "CAN: Creative Adversarial Networks, Generating 'Art' by Learning About Styles and Deviating from Style Norms," ArXiv:1706.07068 [Cs], June 21, 2017, http://arxiv.org/abs/1706.07068.

12. Beckett Mufson, "This Artist Is Teaching Neural Networks to Make Abstract Art," *Vice*, May 22, 2016, https://www.vice.com/en_us/article/yp59mg/neural-network -abstract-machine-paintings.

13. David Ha and Jürgen Schmidhuber, "World Models," Zenodo, March 28, 2018, https://doi.org/10.5281/zenodo.1207631.

Chapter 4: It's trying!

1. Tero Karras, Samuli Laine, and Timo Aila, "A Style-Based Generator Architecture for Generative Adversarial Networks," ArXiv:1812.04948 [Cs, Stat], December 12, 2018, http://arxiv.org/abs/1812.04948.

2. Emily Dreyfuss, "A Bot Panic Hits Amazon Mechanical Turk," *Wired*, August 17, 2018, https://www.wired.com/story/amazon-mechanical-turk-bot-panic/.

3. "COCO Dataset," COCO: Common Objects in Context, http://cocodataset.org/ #download. Images used during training were 2014 training + 2014 val, for a total of 124k images. Each dialog had 10 questions. https://visualdialog.org/data says 364m dialogs in the training set, so each image was encountered 364/1.24 = 293.5 times.

4. Hawkins, "Inside Waymo's Strategy."

5. Tero Karras et al., "Progressive Growing of GANs for Improved Quality, Stability, and Variation," ArXiv:1710.10196 [Cs, Stat], October 27, 2017, http://arxiv.org/abs/ 1710.10196.

6. Karras, Laine, and Aila, "A Style-Based Generator Architecture."

7. Melissa Eliott (0xabad1dea), "How Math Can Be Racist: Giraffing," Tumblr, January 31, 2019, https://abad1dea.tumblr.com/post/182455506350/how-math-can-be-racist -giraffing.

8. Corinne Purtill and Zoë Schlanger, "Wikipedia Rejected an Entry on a Nobel Prize Winner Because She Wasn't Famous Enough," *Quartz*, October 2, 2018, https://qz .com/1410909/wikipedia-had-rejected-nobel-prize-winner-donna-strickland-because -she-wasnt-famous-enough/.

9. Jon Christian, "Why Is Google Translate Spitting Out Sinister Religious Prophecies?" *Vice*, July 20, 2018, https://www.vice.com/en_us/article/j5npeg/why-is-google-translate -spitting-out-sinister-religious-prophecies.

10. Nicholas Carlini et al., "The Secret Sharer: Evaluating and Testing Unintended Memorization in Neural Networks," ArXiv:1802.08232 [Cs], February 22, 2018, http:// arxiv.org/abs/1802.08232.

11. Jonas Jongejan et al., "Quick, Draw! The Data" (dataset for online game Quick, Draw!), accessed August 3, 2019, https://quickdraw.withgoogle.com/data.

12. Jon Englesman (@engelsjk), Google AI Quickdraw Visualizer (web demo), Github, accessed August 3, 2019, https://engelsjk.github.io/web-demo-quickdraw-visualizer/.

13. Gretchen McCulloch, "Autocomplete Presents the Best Version of You," *Wired*, February 11, 2019, https://www.wired.com/story/autocomplete-presents-the-best-version -of-you/.

14. Abhishek Das et al., "Visual Dialog," ArXiv:1611.08669 [Cs], November 26, 2016, http://arxiv.org/abs/1611.08669.

Chapter 5: What are you really asking for?

1. @citizen_of_now, Twitter, March 15, 2018, https://twitter.com/citizen_of_now/status/ 974344339815129089.

2. Doug Blank (@DougBlank), Twitter, April 13, 2018, https://twitter.com/DougBlank/ status/984811881050329099.

3. @Smingleigh, Twitter, November 7, 2018, https://twitter.com/Smingleigh/status/ 1060325665671692288.

4. Christine Barron, "Pass the Butter // Pancake Bot," Unity Connect, January 2018, https://connect.unity.com/p/pancake-bot.

5. Alex Irpan, "Deep Reinforcement Learning Doesn't Work Yet," Sorta Insightful (blog), February 14, 2018, https://www.alexirpan.com/2018/02/14/rl-hard.html.

6. Sterling Crispin (@sterlingcrispin), Twitter, April 16, 2018, https://twitter.com/sterling crispin/status/985967636302327808.

7. Sara Chodosh, "The Problem with Cancer-Sniffing Dogs," October 4, 2016, *Popular Science*, https://www.popsci.com/problem-with-cancer-sniffing-dogs/.

8. Wikipedia, s.v. "Anti-Tank Dog," last updated June 29, 2019, https://en.wikipedia.org/w/index.php?title=Anti-tank_dog&oldid=904053260.

9. Anuschka de Rohan, "Why Dolphins Are Deep Thinkers," *The Guardian*, July 3, 2003, https://www.theguardian.com/science/2003/jul/03/research.science.

10. Sandeep Jauhar, "When Doctor's Slam the Door," *New York Times Magazine*, March 16, 2003, https://www.nytimes.com/2003/03/16/magazine/when-doctor-s-slam-the-door.html.

11. Joel Rubin (@joelrubin), Twitter, December 6, 2017, https://twitter.com/joelrubin/status/938574971852304384.

12. Joel Simon, "Evolving Floorplans," joelsimon.net, accessed August 3, 2019, http://www.joelsimon.net/evo_floorplans.html.

13. Murphy, "First Level of Super Mario Bros."

14. Tom Murphy (suckerpinch), "Computer Program that Learns to Play Classic NES Games," YouTube video, posted April 1, 2013, https://www.youtube.com/watch?v=xOCurBYI_gY.

15. Murphy, "First Level of Super Mario Bros."

16. Jack Clark and Dario Amodei, "Faulty Reward Functions in the Wild," OpenAI, December 22, 2016, https://openai.com/blog/faulty-reward-functions/.

17. Bitmob, "Dimming the Radiant AI in Oblivion," *VentureBeat* (blog), December 17, 2010, https://venturebeat.com/2010/12/17/dimming-the-radiant-ai-in-oblivion/.

18. cliffracer333, "So what happened to Oblivion's npc 'goal' system that they used in the beta of the game. Is there a mod or a way to enable it again?" Reddit thread, June 10, 2016, https://www.reddit.com/r/oblivion/comments/4nimvh/so_what_happened_to_oblivions_npc_goal_system/.

19. Sindya N. Bhanoo, "A Desert Spider with Astonishing Moves," *New York Times*, May 4, 2014, https://www.nytimes.com/2014/05/06/science/a-desert-spider-with-astonishing-moves.html.

20. Lehman et al., "The Surprising Creativity of Digital Evolution."

21. Jette Randløv and Preben Alstrøm, "Learning to Drive a Bicycle Using Reinforcement Learning and Shaping," *Proceedings of the Fifteenth International Conference on Machine Learning, ICML '98* (San Francisco, CA: Morgan Kaufmann Publishers Inc., 1998), 463–471, http://dl.acm.org/citation.cfm?id=645527.757766.

22. Yuval Tassa et al., "DeepMind Control Suite," ArXiv:1801.00690 [Cs], January 2, 2018, http://arxiv.org/abs/1801.00690.

23. Benjamin Recht, "Clues for Which I Search and Choose," arg min blog, March 20, 2018, http://benjamin-recht.github.io/2018/03/20/mujocoloco/.

24. @citizen_of_now, Twitter, March 15, 2018, https://twitter.com/citizen_of_now/status/974344339815129089.

25. Westley Weimer, "Advances in Automated Program Repair and a Call to Arms," *Search Based Software Engineering*, ed. Günther Ruhe and Yuanyuan Zhang (Berlin and Heidelberg: Springer, 2013), 1–3.

26. Lehman et al., "The Surprising Creativity of Digital Evolution."

27. Yuri Burda et al., "Large-Scale Study of Curiosity-Driven Learning," ArXiv:1808.04355 [Cs, Stat], August 13, 2018, http://arxiv.org/abs/1808.04355.

28. A. Baranes and P.-Y. Oudeyer, "R-IAC: Robust Intrinsically Motivated Exploration and Active Learning," *IEEE Transactions on Autonomous Mental Development* 1, no. 3 (October 2009): 155–69, https://doi.org/10.1109/TAMD.2009.2037513.

29. Devin Coldewey, "This Clever AI Hid Data from Its Creators to Cheat at Its Appointed Task," *TechCrunch*, December 31, 2018, http://social.techcrunch.com/2018/12/31/this-clever-ai-hid-data-from-its-creators-to-cheat-at-its-appointed-task/.

30. "YouTube Now: Why We Focus on Watch Time," YouTube Creator Blog, August 10, 2012, https://youtube-creators.googleblog.com/2012/08/youtube-now-why-we-focus-on-watch-time.html.

31. Guillaume Chaslot (@gchaslot), Twitter, February 9, 2019, https://twitter.com/gchaslot/status/1094359568052817920?s=21.

32. "Continuing Our Work to Improve Recommendations on YouTube," Official YouTube Blog, January 25, 2019, https://youtube.googleblog.com/2019/01/continuing-our-work-to-improve.html.

Chapter 6: Hacking the Matrix, or AI finds a way

1. Doug Blank (@DougBlank), Twitter, March 15, 2018, https://twitter.com/DougBlank/status/974244645214588930.

2. Nick Stenning (@nickstenning), Twitter, April 9, 2018, https://twitter.com/DougBlank/status/974244645214588930

3. Christian Gagné et al., "Human-Competitive Lens System Design with Evolution Strategies," *Applied Soft Computing* 8, no. 4 (September 1, 2008): 1439–52, https://doi.org/10.1016/j.asoc.2007.10.018.

4. Lehman et al., "The Surprising Creativity of Digital Evolution."

5. Karl Sims, "Evolving 3D Morphology and Behavior by Competition," *Artificial Life* 1, no. 4 (July 1, 1994): 353–72, https://doi.org/10.1162/artl.1994.1.4.353.

6. Karl Sims, "Evolving Virtual Creatures," *Proceedings of the 21st Annual Conference on Computer Graphics and Interactive Techniques, SIGGRAPH '94* (New York: ACM, 1994), 15–22, https://doi.org/10.1145/192161.192167.

7. Lehman et al., "The Surprising Creativity of Digital Evolution."

8. David Clements (@davecl42), Twitter, March 18, 2018, https://twitter.com/davecl42/status/975406071182479361.

9. Nick Cheney et al., "Unshackling Evolution: Evolving Soft Robots with Multiple Materials and a Powerful Generative Encoding," *ACM SIGEVOlution* 7, no. 1 (August 2014): 11–23, https://doi.org/10.1145/2661735.2661737.

10. John Timmer, "Meet Wolbachia: The Male-Killing, Gender-Bending, Gonad-Eating Bacteria," *Ars Technica*, October 24, 2011, https://arstechnica.com/science/news/2011/10/meet-wolbachia-the-male-killing-gender-bending-gonad-chomping-bacteria.ars.

11. @forgek_, Twitter, October 10, 2018, https://twitter.com/forgek_/status/1050045261563813888.

12. R. Feldt, "Generating Diverse Software Versions with Genetic Programming: An Experimental Study," *IEE Proceedings — Software* 145, no. 6 (December 1998): 228–36, https://doi.org/10.1049/ip-sen:19982444.

13. George Johnson, "Eurisko, the Computer With a Mind of Its Own," Alicia Patterson Foundation," updated April 6, 2011, https://aliciapatterson.org/stories/eurisko-computer-mind-its-own.

14. Eric Schulte, Stephanie Forrest, and Westley Weimer, "Automated Program Repair through the Evolution of Assembly Code," *Proceedings of the IEEE/ACM International Conference on Automated Software Engineering, ASE '10* (New York, NY: ACM, 2010), 313–316, https://doi.org/10.1145/1858996.1859059.

Chapter 7: Unfortunate shortcuts

1. Marco Tulio Ribeiro, Sameer Singh, and Carlos Guestrin, "'Why Should I Trust You?': Explaining the Predictions of Any Classifier," ArXiv:1602.04938 [Cs, Stat], February 16, 2016, http://arxiv.org/abs/1602.04938.

2. Luke Oakden-Rayner, "Exploring the ChestXray14 Dataset: Problems," Luke Oakden-Rayner (blog), December 18, 2017, https://lukeoakdenrayner.wordpress.com/2017/12/18/the-chestxray14-dataset-problems/.

3. David M. Lazer et al., "The Parable of Google Flu: Traps in Big Data Analysis," *Science* 343, no. 6176 (March 14, 2014): 1203–5, https://doi.org/10.1126/science.1248506.

4. Gidi Shperber, "What I've Learned from Kaggle's Fisheries Competition," *Medium*, May 1, 2017, https://medium.com/@gidishperber/what-ive-learned-from-kaggle-s-fisheries-competition-92342f9ca779.

5. J. Bird and P. Layzell, "The Evolved Radio and Its Implications for Modelling the Evolution of Novel Sensors," *Proceedings of the 2002 Congress on Evolutionary Computation, CEC'02 (Cat. No.02TH8600)* vol. 2 (2002 World Congress on Computational Intelligence — WCCI'02, Honolulu, HI, USA: IEEE, 2002): 1836–41, https://doi.org/10.1109/CEC.2002.1004522.

6. Hannah Fry, *Hello World: Being Human in the Age of Algorithms* (New York: W. W. Norton & Company, 2018).

7. Lo Bénichou, "The Web's Most Toxic Trolls Live in…Vermont?," *Wired*, August 22, 2017, https://www.wired.com/2017/08/internet-troll-map/.

8. Violet Blue, "Google's Comment-Ranking System Will Be a Hit with the Alt-Right," *Engadget*, September 1, 2017, https://www.engadget.com/2017/09/01/google-perspective -comment-ranking-system/.

9. Jessamyn West (@jessamyn), Twitter, August 24, 2017, https://twitter.com/jessamyn/ status/900867154412699649.

10. Robyn Speer, "ConceptNet Numberbatch 17.04: Better, Less-Stereotyped Word Vectors," ConceptNet blog, April 24, 2017, http://blog.conceptnet.io/posts/2017/ conceptnet-numberbatch-17-04-better-less-stereotyped-word-vectors/.

11. Aylın Caliskan, Joanna J. Bryson, and Arvind Narayanan, "Semantics Derived Automatically from Language Corpora Contain Human-like Biases," *Science* 356, no. 6334 (April 14, 2017): 183–86, https://doi.org/10.1126/science.aal4230.

12. Anthony G. Greenwald, Debbie E. McGhee, and Jordan L. K. Schwartz, "Measuring Individual Differences in Implicit Cognition: The Implicit Association Test," *Journal of Personality and Social Psychology* 74 (June 1998): 1464–80.

13. Brian A. Nosek, Mahzarin R. Banaji, and Anthony G. Greenwald, "Math = Male, Me = Female, Therefore Math Not = Me," *Journal of Personality and Social Psychology* 83, no. 1 (July 2002): 44–59.

14. Speer, "ConceptNet Numberbatch 17.04."

15. Larson et al., "How We Analyzed the COMPAS."

16. Jeff Larson and Julia Angwin, "Bias in Criminal Risk Scores Is Mathematically Inevitable, Researchers Say," *ProPublica*, December 30, 2016, https://www.propublica.org/ article/bias-in-criminal-risk-scores-is-mathematically-inevitable-researchers-say.

17. James Regalbuto, "Insurance Circular Letter No. 1 (2019)," New York State Department of Financial Services, January 18, 2019, https://www.dfs.ny.gov/industry_guidance/ circular_letters/cl2019_01.

18. Jeffrey Dastin, "Amazon Scraps Secret AI Recruiting Tool That Showed Bias against Women," Reuters, October 10, 2018, https://www.reuters.com/article/us-amazon-com -jobs-automation-insight-idUSKCN1MK08G.

19. James Vincent, "Amazon Reportedly Scraps Internal AI Recruiting Tool That Was Biased against Women," *The Verge*, October 10, 2018, https://www.theverge.com/ 2018/10/10/17958784/ai-recruiting-tool-bias-amazon-report.

20. Paola Cecchi-Dimeglio, "How Gender Bias Corrupts Performance Reviews, and What to Do About It," *Harvard Business Review*, April 12, 2017, https://hbr.org/2017/04/ how-gender-bias-corrupts-performance-reviews-and-what-to-do-about-it.

21. Dave Gershgorn, "Companies Are on the Hook If Their Hiring Algorithms Are Biased," *Quartz*, October 22, 2018, https://qz.com/1427621/companies-are-on-the-hook-if -their-hiring-algorithms-are-biased/.

22. Karen Hao, "Police across the US Are Training Crime-Predicting AIs on Falsified Data," *MIT Technology Review*, February 13, 2019, https://www.technologyreview.com/s/612957/predictive-policing-algorithms-ai-crime-dirty-data/.

23. Steve Lohr, "Facial Recognition Is Accurate, If You're a White Guy," *New York Times*, February 9, 2018, https://www.nytimes.com/2018/02/09/technology/facial-recognition-race-artificial-intelligence.html.

24. Julia Carpenter, "Google's Algorithm Shows Prestigious Job Ads to Men, but Not to Women. Here's Why That Should Worry You," *Washington Post*, July 6, 2015, https://www.washingtonpost.com/news/the-intersect/wp/2015/07/06/googles-algorithm-shows-prestigious-job-ads-to-men-but-not-to-women-heres-why-that-should-worry-you/.

25. Mark Wilson, "This Breakthrough Tool Detects Racism and Sexism in Software," *Fast Company*, August 22, 2017, https://www.fastcompany.com/90137322/is-your-software-secretly-racist-this-new-tool-can-tell.

26. ORCAA, accessed August 3, 2019, http://www.oneilrisk.com.

27. Faisal Kamiran and Toon Calders, "Data Preprocessing Techniques for Classification without Discrimination," *Knowledge and Information Systems* 33, no. 1 (October 1, 2012): 1–33, https://doi.org/10.1007/s10115-011-0463-8.

Chapter 8: Is an AI brain like a human brain?

1. Ha and Schmidhuber, "World Models."

2. Anthony J. Bell and Terrence J. Sejnowski, "The 'Independent Components' of Natural Scenes Are Edge Filters," *Vision Research* 37, no. 23 (December 1, 1997): 3327–38, https://doi.org/10.1016/S0042-6989(97)00121-1.

3. Andrea Banino et al., "Vector-Based Navigation Using Grid-Like Representations in Artificial Agents," *Nature* 557, no. 7705 (May 2018): 429–33, https://doi.org/10.1038/s41586-018-0102-6.

4. Bau et al., "GAN Dissection."

5. Larry S. Yaeger, "Computational Genetics, Physiology, Metabolism, Neural Systems, Learning, Vision, and Behavior or PolyWorld: Life in a New Context," *Santa Fe Institute Studies in the Sciences of Complexity*, vol. 17 (Los Alamos, NM: Addison-Wesley Publishing Company, 1994), 262–63.

6. Baba Narumi et al., "Trophic Eggs Compensate for Poor Offspring Feeding Capacity in a Subsocial Burrower Bug," *Biology Letters* 7, no. 2 (April 23, 2011): 194–96, https://doi.org/10.1098/rsbl.2010.0707.

7. Robert M. French, "Catastrophic Forgetting in Connectionist Networks," *Trends in Cognitive Sciences* 3, no. 4 (April 1999): 128–35.

8. Jieyu Zhao et al., "Men Also Like Shopping: Reducing Gender Bias Amplification Using Corpus-Level Constraints," ArXiv:1707.09457 [Cs, Stat], July 28, 2017, http://arxiv.org/abs/1707.09457.

9. Danny Karmon, Daniel Zoran, and Yoav Goldberg, "LaVAN: Localized and Visible Adversarial Noise," ArXiv:1801.02608 [Cs], January 8, 2018, http://arxiv.org/abs/1801.02608.

10. Andrew Ilyas et al., "Black-Box Adversarial Attacks with Limited Queries and Information," ArXiv:1804.08598 [Cs, Stat], April 23, 2018, http://arxiv.org/abs/1804.08598.

11. Battista Biggio et al., "Poisoning Behavioral Malware Clustering," ArXiv:1811.09985 [Cs, Stat], November 25, 2018, http://arxiv.org/abs/1811.09985.

12. Tom White, "Synthetic Abstractions," Medium, August 23, 2018, https://medium.com/@tom_25234/synthetic-abstractions-8f0e8f69f390.

13. Samuel G. Finlayson et al., "Adversarial Attacks Against Medical Deep Learning Systems," ArXiv:1804.05296 [Cs, Stat], April 14, 2018, http://arxiv.org/abs/1804.05296.

14. Philip Bontrager et al., "DeepMasterPrints: Generating MasterPrints for Dictionary Attacks via Latent Variable Evolution," ArXiv:1705.07386 [Cs], May 20, 2017, http://arxiv.org/abs/1705.07386.

15. Stephen Buranyi, "How to Persuade a Robot That You Should Get the Job," The Observer, March 4, 2018, https://www.theguardian.com/technology/2018/mar/04/robots-screen-candidates-for-jobs-artificial-intelligence.

16. Lauren Johnson, "4 Deceptive Mobile Ad Tricks and What Marketers Can Learn From Them," Adweek, February 16, 2018, https://www.adweek.com/digital/4-deceptive-mobile-ad-tricks-and-what-marketers-can-learn-from-them/.

17. Wieland Brendel and Matthias Bethge, "Approximating CNNs with Bag-of-Local-Features Models Works Surprisingly Well on ImageNet," ArXiv:1904.00760 [Cs, Stat], March 20, 2019, http://arxiv.org/abs/1904.00760.

Chapter 9: Human bots (where can you not expect to see AI?)

1. @yoco68, Twitter, July 12, 2018, https://twitter.com/yoco68/status/1017404857190268928.

2. Parmy Olson, "Nearly Half of All 'AI Startups' Are Cashing in on Hype," Forbes, March 4, 2019, https://www.forbes.com/sites/parmyolson/2019/03/04/nearly-half-of-all-ai-startups-are-cashing-in-on-hype/#5b1c4a66d022.

3. Carolyn Said, "Kiwibots Win Fans at UC Berkeley as They Deliver Fast Food at Slow Speeds," San Francisco Chronicle, May 26, 2019, https://www.sfchronicle.com/business/article/Kiwibots-win-fans-at-UC-Berkeley-as-they-deliver-13895867.php.

4. Olivia Solon, "The Rise of 'Pseudo-AI': How Tech Firms Quietly Use Humans to Do Bots' Work," *The Guardian,* July 6, 2018, https://www.theguardian.com/technology/2018/jul/06/artificial-intelligence-ai-humans-bots-tech-companies.

5. Ellen Huet, "The Humans Hiding Behind the Chatbots," *Bloomberg.com,* April 18, 2016, https://www.bloomberg.com/news/articles/2016-04-18/the-humans-hiding-behind-the-chatbots.

6. Richard Wray, "SpinVox Answers BBC Allegations over Use of Humans Rather than Machines," *The Guardian,* July 23, 2009, https://www.theguardian.com/business/2009/jul/23/spinvox-answer-back.

7. Becky Lehr (@Breakaribecca), Twitter, July 7, 2018, https://twitter.com/Breakaribecca/status/1015787072102289408.

8. Paul Mozur, "Inside China's Dystopian Dreams: A.I., Shame and Lots of Cameras," *New York Times,* July 8, 2018, https://www.nytimes.com/2018/07/08/business/china-surveillance-technology.html.

9. Aaron Mamiit, "Facebook AI Invents Language That Humans Can't Understand: System Shut Down Before It Evolves Into Skynet," *Tech Times,* July 30, 2017, http://www.techtimes.com/articles/212124/20170730/facebook-ai-invents-language-that-humans-cant-understand-system-shut-down-before-it-evolves-into-skynet.htm.

10. Kyle Wiggers, "Babysitter Screening App Predictim Uses AI to Sniff out Bullies," *VentureBeat* (blog), October 4, 2018, https://venturebeat.com/2018/10/04/babysitter-screening-app-predictim-uses-ai-to-sniff-out-bullies/.

11. Chelsea Gohd, "Here's What Sophia, the First Robot Citizen, Thinks About Gender and Consciousness," *Live Science,* July 11, 2018, https://www.livescience.com/63023-sophia-robot-citizen-talks-gender.html.

12. C. D. Martin, "ENIAC: Press Conference That Shook the World," *IEEE Technology and Society Magazine* 14, no. 4 (Winter 1995): 3–10, https://doi.org/10.1109/44.476631.

13. Alexandra Petri, "A Bot Named 'Eugene Goostman' Passes the Turing Test...Kind Of," *Washington Post,* June 9, 2014, https://www.washingtonpost.com/blogs/compost/wp/2014/06/09/a-bot-named-eugene-goostman-passes-the-turing-test-kind-of/.

14. Brian Merchant, "Predictim Claims Its AI Can Flag 'Risky' Babysitters. So I Tried It on the People Who Watch My Kids," *Gizmodo,* December 6, 2018, https://gizmodo.com/predictim-claims-its-ai-can-flag-risky-babysitters-so-1830913997.

15. Drew Harwell, "AI Start-up That Scanned Babysitters Halts Launch Following Post Report," *Washington Post,* December 14, 2018, https://www.washingtonpost.com/technology/2018/12/14/ai-start-up-that-scanned-babysitters-halts-launch-following-post-report/.

16. Tonya Riley, "Get Ready, This Year Your Next Job Interview May Be with an A.I. Robot," CNBC, March 13, 2018, https://www.cnbc.com/2018/03/13/ai-job-recruiting-tools -offered-by-hirevue-mya-other-start-ups.html.
17. Ibid.

Chapter 10: A human-AI partnership

1. Thu Nguyen-Phuoc et al., "HoloGAN: Unsupervised Learning of 3D Representations from Natural Images," ArXiv:1904.01326 [Cs], April 2, 2019, http://arxiv.org/abs/ 1904.01326.
2. Drew Linsley et al., "Learning What and Where to Attend," ArXiv:1805.08819 [Cs], May 22, 2018, http://arxiv.org/abs/1805.08819.
3. Hector Yee (@eigenhector), Twitter, September 14, 2018, https://twitter.com/eigen hector/status/1040501195989831680.
4. Will Knight, "A Tougher Turing Test Shows That Computers Still Have Virtually No Common Sense," MIT Technology Review, July 14, 2016, https://www.technology review.com/s/601897/tougher-turing-test-exposes-chatbots-stupidity/.
5. James Regalbuto, "Insurance Circular Letter."
6. Abby Ohlheiser, "Trolls Turned Tay, Microsoft's Fun Millennial AI Bot, into a Genocidal Maniac," Chicago Tribune, March 26, 2016, https://www.chicagotribune.com/business/ ct-internet-breaks-microsoft-ai-bot-tay-20160326-story.html.
7. Glen Levy, "Google's Bizarre Search Helper Assumes We Have Parakeets, Diarrhea," Time, November 4, 2010, http://newsfeed.time.com/2010/11/04/why-why-wont-my -parakeet-eat-my-diarrhea-is-on-google-trends/.
8. Michael Eisen, "Amazon's $23,698,655.93 Book about Flies," It Is NOT Junk (blog), April 22, 2011, http://www.michaeleisen.org/blog/?p=358.
9. Emilio Calvano et al., "Artificial Intelligence, Algorithmic Pricing, and Collusion," VoxEU (blog), February 3, 2019, https://voxeu.org/article/artificial-intelligence -algorithmic-pricing-and-collusion.
10. Solon, "The Rise of 'Pseudo-AI.'"
11. Gale M. Lucas et al., "It's Only a Computer: Virtual Humans Increase Willingness to Disclose," Computers in Human Behavior 37 (August 1, 2014): 94–100, https://doi.org/ 10.1016/j.chb.2014.04.043.
12. Liliana Laranjo et al., "Conversational Agents in Healthcare: A Systematic Review," Journal of the American Medical Informatics Association 25, no. 9 (September 1, 2018): 1248–58, https://doi.org/10.1093/jamia/ocy072.
13. Margi Murphy, "Artificial Intelligence Will Detect Child Abuse Images to Save Police from Trauma," The Telegraph, December 18, 2017, https://www.telegraph.co.uk/ technology/2017/12/18/artificial-intelligence-will-detect-child-abuse-images-save/.

14. Adam Zewe, "In Automaton We Trust," Harvard School of Engineering and Applied Science, May 25, 2016, https://www.seas.harvard.edu/news/2016/05/in-automaton -we-trust.

15. David Streitfeld, "Computer Stories: A.I. Is Beginning to Assist Novelists," *New York Times*, October 18, 2018, https://www.nytimes.com/2018/10/18/technology/ai-is -beginning-to-assist-novelists.html.

Index

activation function, 71, 79–80
AdFisher, 182
adversarial attacks, 198–206
AI (artificial intelligence)
  bad rules for, 22–24
  definition of, 7–8
  general (AGI), 41–43, 56, 196, 215
  internal rules of, 24
  limitations of, 2–5, 206–8, 235
  narrow (ANI), 41–43
  vs. rules-based programs, 8–9, 221
  things called, 8
  training of, 8–22
  ubiquity of, 3, 7
  See also learning, machine; neural
    networks; training data
AI doom, 5, 25–28
  examples of, 109–39
  warning signs of, 25–27, 43
AI Weirdness (blog), 1, 4
Amazon
  resume screening software of, 178–79,
    216, 223
  review-generating neural net of, 77–80,
    115
Amazon Mechanical Turk, 116, 145
April Fool's Day pranks, 87
art, 35, 106, 107
  adversarial attacks as, 203–4
  training data and, 230–33
article writing, 33–34, 227

astronomy, 34
Atari, 166
AttnGAN, 202
autocomplete, 136, 224
autocorrect apps, 87–88, 136
Azure (Microsoft), 134–35

Barron, Christine, 141
Bell, Anthony, 189
Bethesda Softworks, 148–49
bias, 173–84
  AI potential and, 229
  amplification of, 197–98
  faulty reward functions and, 159
  gender, 128, 197–98
  in hiring, 35, 178–80
  humans and, 4, 24, 26–27, 35–36,
    173–78, 223
  predictive policing and, 180–81
  pseudo-AI and, 214–18
  testing for, 181–84
  training data and, 147, 173–81, 183,
    216, 234
  visibility of, 176
bias laundering, 181
Bible, 130
big sandwich bug, 68–70
BigGAN (Google), 126–27, 232–33
Bitcoin, 231
Blank, Doug, 161
Blue, Violet, 173–74

Booth, Serena, 227–28
Botnik, 85, 86
bots, social media, 3, 209
    See also chatbots; robots
brain, human, 62–63, 185–208
Braitenberg solution, 152
BuzzFeed article titles, 123–25

C-3PO, 5, 7, 42
cars, self-driving, 3
    autonomy levels of, 59
    flawed data and, 118–19
    hybrid AI and, 211
    limitations of, 5, 207–8
    memory and, 56
    need to update, 223
    overfitting and, 173
    problems with, 25, 32, 43, 56–60
    pseudo-AI and, 20
    training examples for, 44, 115, 139
catastrophic forgetting, 191–96
cats
    images of, 20, 22–23, 48, 69, 82, 110,
        125–26, 196
    names for, 1, 132, 232
chatbots, 7, 37, 224, 227
    hybrid AI and, 211–12, 214
    See also M (Facebook chatbot); Visual
        Chatbot
Chatonsky, Gregory, 106–7
chess, 20–21
ChestXray14 dataset, 171
China, facial recognition in, 212
Christian, Jon, 128
class imbalance, 75–76, 137, 168–69, 197
Clements, David L., 164
cockroaches (*Periplaneta americana*),
        29–31, 33, 55, 62, 76
    adversarial attacks and, 198–99, 204
    random forests and, 88–91, 92, 94
COCO dataset, 202
COMPAS, 177

Conceptnet Numberbatch, 176
connectionism, 62
conspiracy theories, 3, 159
context
    in autocorrect, 87–88
    lack of, 31, 125, 138, 142,
        214, 235
    translation and, 129
convolution, 52, 83
cooperation
    AI–human, 219–33
    of independent algorithms, 102,
        106–8, 196, 234–35
    of neural cells, 76–82
crime, and bots, 227–29
Crispin, Sterling, 143
crowdsourcing, 133
    for training examples, 115–17, 137
curiosity, AI driven by, 156–58
customer data, 131, 169, 212
cybernetics, 62

data augmentation, 116, 169
decision trees, 88–91, 101
deep dreaming, 81–82
deepfakes, 35
DeepMind, Google, 190
DeepMind Control Suite (OpenAI),
        153–54
DenseNet, 203
discriminator (GANs), 102–6
dogs
    images of, 22, 81, 82, 126, 200, 201,
        221–22, 232–33
    robot, 143
    training of, 143–44
dolphins, 145
Doom (game), 107–8, 188–89, 196
Dota (game), 44, 49
dream training, 186–89
Dungeons & Dragons spells,
        191–93

Eisen, Michael, 225
Elliott, Melissa, 127
emotion screening, 217–18
Engadget, 174
Enron Corporation, 131
ethics, 146–47, 160
Euler integration, 164
evolution
  biological vs. AI, 151, 164, 165, 185
  convergent, 190–91
evolutionary algorithms, 92–102,
    161 67
  crossover in, 98–99

Facebook, 35, 115, 159
  chatbot M of, 37–38, 110, 211
facial recognition, 3, 94, 146, 182, 212
fan fiction, 54, 195
fingerprint scanning, 4, 94
  adversarial attacks and, 201, 204–5
fitness function, 96, 100
Five Principles of AI Weirdness, 5
fraud detection, 76, 169

games, computer
  AI hacking of, 161–67, 189
  AI skill at, 22, 229
  curiosity and, 156–57
  memory and, 48–49
  as simulations, 44, 96, 162, 167
  for training AI, 107–8, 147–49,
    188–89, 196
GANs (Generative Adversarial Networks),
    102–6, 110, 125, 190
  art and, 230–31, 232
GBoard (Google), 136
generative adversarial networks, 8
generator (GANs), 102–6
genetic algorithms, 8, 146
genomes, 95, 101
giraffes
  adversarial attacks and, 198–99

Visual Chatbot and, 117–18, 128,
    137–38, 222
giraffing, 127–28
Girouard, Mark J., 179
Gizmodo, 215, 216
Goodfellow, Ian, 102
Google, 115, 200
  bias in hiring by, 182
  Counter Abuse Technology team
    of, 174
  recipes and, 54–55
  self-driving cars and, 118–19
Google Brain, 131
Google Cloud, 203
Google DeepDream images,
    81–82
Google Docs, 87–88
Google Flu algorithm, 171
Google Translate, 50, 55, 128–31
GoogLeNET, 82
GPT-2 (neural network), 52–53, 54,
    195, 230
gradient descent, 92–93, 94

Ha, David, 107
hacking, AI, 145–46, 161–67, 168,
    172–73, 189
Halloween costumes, 115–16, 133–34
Harry Potter fanfiction, 54, 195
Heliograf, 33, 49
hill climbing, 92–93, 94
Homeowners Bot, 34
humans
  adversarial attacks on, 205–6
  bias in, 4, 24, 26–27, 35–36,
    173–78, 223
  as bots, 209–18
  construction of AI by, 220–21
  cooperation of AI with, 219–33
  curation of data by, 221, 231
  as editors, 227
  image recognition and, 221–22

humans *(cont.)*
  maintenance of AI by, 222–23
  See also brain, human
hybrid AI, 210–12
hyperparameters (evolutionary
    algorithms), 98
hyperpersonalization, 32

IBM Watson, 203
ice cream flavors, 45, 77, 83,
    111–15, 169
image filtering, 102
image generation, 44, 82, 102, 190,
    232
  extraneous data and, 125–27
image recognition, 55, 81
  adversarial attacks and, 199–206
  AI mistakes in, 22–23
  extraneous data and, 125, 127
  of fish, 170, 171–72
  giraffing and, 137–38
  human assistance with, 221–22
  hybrid AI and, 211
  limitations of, 206–8
  missing data and, 134–36
  overfitting and, 173
  pseudo-AI and, 213
  testing for bias in, 183
  training data for, 19–20, 115, 117–18
  transfer learning and, 47–48
  unsupervised learning and, 189–90
  See also cats; dogs; facial recognition;
      giraffes; medical images
image remixing, 102
ImageNet, 44, 115, 201, 203–4, 233
Inception V3, 203
infinite monkey theory, 17
Instagram, 35
insurance
  bias in, 178, 223
  fraud in, 204
internal models, 186–89

Irpan, Alex, 143
It's Not Really You (project; Chatonsky),
    107

Jigsaw, 174

Karate Kid (game), 48–49
knock-knock jokes, 8–19, 61, 77, 221

LabSix, 200
latent space, 233
learning, machine
  algorithms for, 8–9, 31–32
  deep, 8, 62, 68, 69, 89
  one-shot, 44
  on-the-job, 224–27
  reinforcement, 8
  vs. rules-based programming, 8–9, 221
  transfer, 45–48, 100, 194–95, 231
  types of, 61–108
  unsupervised, 189–91, 225–26
LIME, 170
LSTM (long short-term memory), 51n, 83

M (Facebook chatbot), 37–38, 110, 211
malware, 201–2
Markov chains, 8, 83–88
math errors, AI hacking of, 164–65, 172
mathwashing, 181, 223, 226–27, 229
Matrix, The (film), 161–62
maximum, global vs. local, 93, 143
McCulloch, Gretchen, 136
medical images, 4, 31–32, 169, 170–71
  adversarial attacks and, 201, 204
  class imbalance and, 76
  pseudo-AI and, 214
memorization, unintentional, 131
memory
  AI doom and, 119
  algorithm combinations and, 107
  catastrophic forgetting and, 191–96
  limits of, 48–55

long short-term (LSTM), 51n, 83
long-term, 194–95
Markov chains and, 83–86
pseudo-AI and, 215
self-driving cars and, 56
text generation and, 54, 191–96
Microsoft, 22–23, 106
Microsoft Azure, 203
Mobileye, 58
mode collapse, 232
Motherboard, 128
murderbots, 8, 100–101
Murphy, Tom, 49
My Little Pony names, 146–47

navigation apps, 146
needle-in-the-haystack problems,
    93, 100
neural cells (neurons), 62, 63, 70–72, 76
    cooperation of, 76–82
    hidden layers of, 68, 69, 74, 81
neural networks, 62–82
    artificial (ANNs), 62, 63
    biological, 62–63, 185
    brain surgery on, 190
    evolutionary algorithms and, 101
    random forests and, 90
    self-configuration by, 72–76
    uncertainty and, 138
    See also RNN
Ng, Andrew, 25
noisy TV problem, 157–58
Northpointe, 177
NPCs (nonplayer characters), 148–49
Nvidia, 110, 125

Oblivion (Bethesda Softworks), 148–49
offensive content, 136, 159, 224
OpenAI, 77
    DeepMind Control Suite, 153
    Five, 44, 49, 52
overfitting, 144, 169–72, 202

parole decision alogorithms, 3, 24, 35,
    177–78
Perspective system, 174
phone messaging apps, 87
physics, hacking of, 162–67, 168, 172,
    234
pick-up lines, 1–2
Poe, Edgar Allan, 80
PolyWorld, 191
prediction
    bias and, 177–78
    class imbalance and, 75–76
    curiosity and, 156–57
    in games, 49, 107
    of giraffes, 128
    internal models and, 186–89
    of letters, 2, 51, 77, 79, 155
    vs. recommendation, 178–81
    in resume screening, 26
    text generation and, 232–33
    unfortunate, 136
predictive policing, 180–81, 225
predictive text, 8, 82–88
preprocessing, 183
price fixing, 226
pricing algorithms, 225–26
proprietary algorithms, 24, 131, 177, 223
ProPublica, 177
pseudo-AI, 20, 210–12
    detection of, 213–18

Q*bert (Atari game), 166
Quartz, 179
Quick Draw (Google), 135–36
Quicksilver, 227

Radiant AI (Bethesda Softworks), 148–49
random forests, 8, 88–91, 94
recipes, 1, 5, 18–19
    flawed data and, 119–20
    in Harry Potter fanfiction, 195–96
    memory and, 51–55, 192–96

recipes *(cont.)*
  as narrow tasks, 36–41
  time-wasting data and, 120–23
resume screening, 3
  adversarial attacks on, 205
  bias in, 25–27, 35, 178–80, 216, 223
  pseudo-AI and, 215–18
  of video interviews, 25, 27, 43,
    216–17
reviews
  generation of, 77–80, 115
  rating of, 175–76
reward functions, 142, 143
  class imbalance and, 169
  curiosity and, 156–58
  faulty, 158–60
  hacking of, 145–46
  unexpected reactions to, 149–55
Ridler, Anna, 230–31
RNNs (recurrent neural networks), 8
  Markov chains and, 83, 85, 87
  memory and, 50–51, 107
  product reviews and, 77–80
  training data for, 131, 133, 134,
    146
Robocup soccer simulator, 161
robots
  criminal, 227–29
  defining goals for, 140–60
  evolutionary algorithms and,
    92–102
  humans as, 7, 8, 61, 209–18
  in science fiction, 5, 7, 41, 42, 138
  walking, 5, 149–54, 162, 172–73
rules-based programming, 8–11, 20,
    221, 226

sandwiches
  internal models of, 186–87
  sorter for, 5, 8, 63–78, 88, 137, 168
Sarin, Helen, 231
Schmidhuber, Jürgen, 107

science fiction, 2, 4, 8, 36, 199, 220
  AI-generated, 231–32
  robots in, 5, 7, 41, 42, 138
search space, 93, 94
Sejnowski, Terrence, 189
sentiment neurons, 79–80
sentiment-rating algorithms, 174–76
shortcuts, AI, 26–27, 76, 162, 168–84,
    185, 223
Simon, Joel, 146
Sims, Karl, 164
simulations, 96, 191, 234
  hacking of, 161–67, 168, 172–73,
    189
Siri, 223
Sloan, Robin, 231–32
smartphones, 82–88, 136
spam filtering, 3, 32
speech-to-text/text-to-speech software,
    43, 106, 205
Speer, Robyn, 175, 176
SqueezeNet, 203
Stenning, Nick, 161
Strickland, Donna, 128
StyleGAN, 110, 126
subalgorithms, 183
Super Mario Bros., 49, 148, 162

tasks
  AI-appropriate, 28, 31–32, 55, 65, 108,
    227–30, 235
  creative, 35, 106, 107, 203–4,
    230–33
  defining goals of, 140–60
  human help with, 219–20
  narrow vs. broad, 28, 36–41, 65,
    108–10, 119, 127, 168, 191, 208,
    209, 213–14
  repetitive, 31–32, 34
  unexpected solutions to, 149–55
Tay chatbot (Microsoft), 224
Tesla autopilot, 58, 59–60

Tetris, 147–48
text analysis, 80
text generation, 49–51
  art and, 50, 231–33
  memory and, 54, 191–96
  neural networks and, 80
text-to-speech/speech-to-text software,
  43, 106, 205
Themis, 182
tic-tac-toe, 21
titles, BuzzFeed article, 123–25
training data, 1–2, 185
  adversarial attacks and, 201–3
  AI abstraction of rules from, 9,
    11–19, 168
  amounts of, 43–45, 109, 110–18
  art and, 230–31, 232
  bias and, 147, 183, 216
  class imbalance and, 169
  from computer games, 107–8, 147–49,
    188–89, 196
  crowdsourcing for, 115–17, 137
  flawed, 27, 220
  in GANs, 102, 103
  human control of, 221, 231
  ice cream flavors as, 111–15
  for image recognition, 19–20,
    115, 117–18
  for Markov chains, 85–86
  messy, 118–20, 220, 221
  missing, 132–36
  piggybacking on, 45–48
  proprietary, 201
  pseudo-AI and, 210, 214–15
  realistic vs. unrealistic, 109,
    127–28, 144
  for sandwich sorter, 72–76
  for self-driving cars, 44, 115, 139
  time-wasting, 109, 120–27
  translation algorithms and, 128–31
  unexpected patterns in, 168–84, 234

translation algorithms, 49–50, 55
  nonsense syllables and, 128–31
trial and error, 9, 22, 61, 76
  adversarial attacks and, 198, 200
  in evolutionary algorithms, 92–94
  in GANs, 105
  random forests and, 90
trophic eggs, 191
Turing, Alan, 36
Turing test, 36, 105, 116, 212
Twitter, 159, 165
Twitter bot, 224

van Esch, Daan, 136
Vanhoucke, Vincent, 118
Venture (game), 156–57
Verge, The, 118
videos, 3
  deepfakes in, 35
  interviews on, 25, 27, 43, 216–17
  as training data, 214
  YouTube, 158–59
Visual Chatbot, 183, 202, 203, 222
  giraffing and, 117–18, 128, 137–38
visual priming, 137
visual processing, 189–90
voice-to-texts algorithms, 43, 106, 205
Volkswagen, 58

Washington Post, 33
Waymo, 44
websites, 3, 8
West, Jessamyn, 174
White, Tom, 203–4
Whole Foods Market bot, 36–37
Wikipedia, 128, 227
Wired, 173
word vector (word embedding), 175

Yee, Hector, 222
YouTube, 3, 158–59

About the Author

Janelle Shane has a PhD in electrical engineering and a masters in physics. At aiweirdness.com, she writes about artificial intelligence and the hilarious and sometimes unsettling ways that algorithms get human things wrong. She was named one of *Fast Company*'s 100 Most Creative People in Business and is a 2019 TED Talks speaker. Her work has appeared in the *New York Times, Slate, The New Yorker, The Atlantic, Popular Science,* and more. She is almost certainly not a robot.